GENOMICS-ENABLED LEARNING HEALTH CARE SYSTEMS
Gathering and Using Genomic Information to
Improve Patient Care and Research

WORKSHOP SUMMARY

Roundtable on Translating Genomic-Based Research
for Health

Board on Health Sciences Policy

Sarah H. Beachy, Steve Olson, and Adam C. Berger, *Rapporteurs*

INSTITUTE OF MEDICINE
OF THE NATIONAL ACADEMIES

THE NATIONAL ACADEMIES PRESS
Washington, D.C.
www.nap.edu

THE NATIONAL ACADEMIES PRESS • 500 Fifth Street, NW • Washington, DC 20001

NOTICE: The workshop that is the subject of this workshop summary was approved by the Governing Board of the National Research Council, whose members are drawn from the councils of the National Academy of Sciences, the National Academy of Engineering, and the Institute of Medicine.

This project was supported by contracts between the National Academy of Sciences and the American Academy of Nursing (unnumbered contract); American College of Medical Genetics and Genomics (unnumbered contract); American Heart Association (unnumbered contract); American Medical Association (unnumbered contract); American Society of Human Genetics (unnumbered contract); Association for Molecular Pathology (unnumbered contract); Biogen Idec (unnumbered contract); Blue Cross and Blue Shield Association (unnumbered contract); College of American Pathologists (unnumbered contract); Department of Veterans Affairs (Contract No. VA240-14-C-0037); Eli Lilly and Company (unnumbered contract); Genetic Alliance (unnumbered contract); Health Resources and Services Administration (Contract No. HHSH250200976014I, Order No. HHSH25034021T); International Society for Cardiovascular Translational Research (unnumbered contract); Janssen Research & Development, LLC (unnumbered contract); Kaiser Permanente Program Offices Community Benefit II at the East Bay Community Foundation (Contract No. 20121257); Merck & Co., Inc. (Contract No. CMO-140505-000393); National Cancer Institute (Contract No. HHSN263201200074I, TO#5); National Human Genome Research Institute (Contract No. HHSN263201200074I, TO#5); National Institute of Mental Health (Contract No. HHSN263201200074I, TO#5); National Institute of Nursing Research (Contract No. HHSN263201200074I, TO#5); National Institute on Aging (Contract No. HHSN263201200074I, TO#5); National Society of Genetic Counselors (unnumbered contract); Northrop Grumman Health IT (unnumbered contract); Pfizer Inc. (unnumbered contract); and PhRMA (unnumbered contract). The views presented in this publication do not necessarily reflect the views of the organizations or agencies that provided support for the activity.

International Standard Book Number-13: 978-0-309-37112-4
International Standard Book Number-10: 0-309-37112-0

Additional copies of this workshop summary are available for sale from the National Academies Press, 500 Fifth Street, NW, Keck 360, Washington, DC 20001; (800) 624-6242 or (202) 334-3313; http://www.nap.edu.

For more information about the Institute of Medicine, visit the IOM home page at: **www.iom.edu.**

Copyright 2015 by the National Academy of Sciences. All rights reserved.

Printed in the United States of America

The serpent has been a symbol of long life, healing, and knowledge among almost all cultures and religions since the beginning of recorded history. The serpent adopted as a logotype by the Institute of Medicine is a relief carving from ancient Greece, now held by the Staatliche Museen in Berlin.

Suggested citation: IOM (Institute of Medicine). 2015. *Genomics-enabled learning health care systems: Gathering and using genomic information to improve patient care and research: Workshop summary.* Washington, DC: The National Academies Press.

*"Knowing is not enough; we must apply.
Willing is not enough; we must do."*
—Goethe

INSTITUTE OF MEDICINE
OF THE NATIONAL ACADEMIES

Advising the Nation. Improving Health.

THE NATIONAL ACADEMIES
Advisers to the Nation on Science, Engineering, and Medicine

The **National Academy of Sciences** is a private, nonprofit, self-perpetuating society of distinguished scholars engaged in scientific and engineering research, dedicated to the furtherance of science and technology and to their use for the general welfare. Upon the authority of the charter granted to it by the Congress in 1863, the Academy has a mandate that requires it to advise the federal government on scientific and technical matters. Dr. Ralph J. Cicerone is president of the National Academy of Sciences.

The **National Academy of Engineering** was established in 1964, under the charter of the National Academy of Sciences, as a parallel organization of outstanding engineers. It is autonomous in its administration and in the selection of its members, sharing with the National Academy of Sciences the responsibility for advising the federal government. The National Academy of Engineering also sponsors engineering programs aimed at meeting national needs, encourages education and research, and recognizes the superior achievements of engineers. Dr. C. D. Mote, Jr., is president of the National Academy of Engineering.

The **Institute of Medicine** was established in 1970 by the National Academy of Sciences to secure the services of eminent members of appropriate professions in the examination of policy matters pertaining to the health of the public. The Institute acts under the responsibility given to the National Academy of Sciences by its congressional charter to be an adviser to the federal government and, upon its own initiative, to identify issues of medical care, research, and education. Dr. Victor J. Dzau is president of the Institute of Medicine.

The **National Research Council** was organized by the National Academy of Sciences in 1916 to associate the broad community of science and technology with the Academy's purposes of furthering knowledge and advising the federal government. Functioning in accordance with general policies determined by the Academy, the Council has become the principal operating agency of both the National Academy of Sciences and the National Academy of Engineering in providing services to the government, the public, and the scientific and engineering communities. The Council is administered jointly by both Academies and the Institute of Medicine. Dr. Ralph J. Cicerone and Dr. C. D. Mote, Jr., are chair and vice chair, respectively, of the National Research Council.

www.national-academies.org

PLANNING COMMITTEE[1]

GEOFFREY GINSBURG (*Co-Chair*), Director, Center for Applied Genomics and Precision Medicine, Professor of Medicine and of Pathology and Biomedical Engineering, Duke University
SAM SHEKAR (*Co-Chair*), Chief Medical Officer, Northrop Grumman Health IT
JENNIFER HALL, Director, Program in Translational Genomics, Lillehei Heart Institute, Associate Professor of Medicine, University of Minnesota
ANDREW KASARSKIS, Vice Chair, Department of Genetics and Genomic Sciences, Co-Director, Institute for Genomics and Multiscale Biology, Icahn School of Medicine at Mount Sinai
DEBRA LEONARD, Professor and Chair of Pathology and Laboratory Medicine, The University of Vermont Medical Center
JAMES O'LEARY, Chief Innovation Officer, Genetic Alliance
MICHELLE ANN PENNY, Senior Director, TTx Genetics and Bioinformatics, Eli Lilly and Company
RONALD PRZYGODZKI, Acting Director, Biomedical Laboratory Research and Development, Associate Director for Genomic Medicine, Office of Research and Development, Department of Veterans Affairs
MICHAEL S. WATSON, Executive Director, American College of Medical Genetics and Genomics
CATHERINE A. WICKLUND, Director, Graduate Program in Genetic Counseling, Associate Professor, Department of Obstetrics and Gynecology, Northwestern University
JANET K. WILLIAMS, Professor of Nursing, Chair of the Social Science and Behavioral Research Institutional Review Board, University of Iowa

---

[1]The planning committee's role was limited to planning the workshop. The workshop summary has been prepared by the rapporteurs as a factual account of what occurred at the workshop. Statements, recommendations, and opinions expressed are those of individual presenters and participants and are not necessarily endorsed or verified by the Institute of Medicine. They should not be construed as reflecting any group consensus.

IOM Staff

ADAM C. BERGER, Project Director
SARAH H. BEACHY, Associate Program Officer
MEREDITH HACKMANN, Senior Program Assistant

ROUNDTABLE ON TRANSLATING GENOMIC-BASED RESEARCH FOR HEALTH[1]

GEOFFREY GINSBURG (*Co-Chair*), Director, Center for Applied Genomics and Precision Medicine, Duke University, Durham, NC
SHARON TERRY (*Co-Chair*), President and Chief Executive Officer, Genetic Alliance, Washington, DC
NAOMI ARONSON, Executive Director, Technology Evaluation Center, Blue Cross and Blue Shield Association, Chicago, IL
EUAN ANGUS ASHLEY, Representative of the American Heart Association; Director, Center for Inherited Cardiovascular Disease, Stanford University School of Medicine, Palo Alto, CA
PAUL R. BILLINGS, former Chief Medical Officer, Life Technologies, Carlsbad, CA
BRUCE BLUMBERG, Institutional Director of Graduate Medical Education, Northern California Kaiser Permanente, The Permanente Medical Group, Oakland, CA
PAMELA BRADLEY (*until October 2014*), Staff Fellow, Personalized Medicine Staff, Office of In Vitro Diagnostics and Radiological Health, Center for Devices and Radiological Health, U.S. Food and Drug Administration, Silver Spring, MD
PHILIP J. BROOKS, Health Scientist Administrator, Office of Rare Diseases Research, National Center for Advancing Translational Sciences, National Institutes of Health, Bethesda, MD
JOHN CARULLI, Director, Translational Genomics, Biogen Idec, Cambridge, MA
ANN CASHION, Scientific Director, National Institute of Nursing Research, National Institutes of Health, Bethesda, MD
ROBERT B. DARNELL, President and Scientific Director, New York Genome Center; Investigator, Howard Hughes Medical Institute, Heilbrunn Cancer Professor and Senior Physician, Head, Laboratory of Molecular Neuro-Oncology, Rockefeller University, New York, NY
MICHAEL J. DOUGHERTY, Director of Education, American Society of Human Genetics, Bethesda, MD
W. GREGORY FEERO, Contributing Editor, *Journal of the American Medical Association*, Chicago, IL

---

[1] Institute of Medicine forums and roundtables do not issue, review, or approve individual documents. The responsibility for the published workshop summary rests with the workshop rapporteurs and the institution.

ANDREW N. FREEDMAN, Branch Chief, Clinical and Translational Epidemiology Branch, Epidemiology and Genetics Research Program, Division of Cancer Control and Population Sciences, National Cancer Institute, Rockville, MD

JENNIFER L. HALL, Representative of the International Society for Cardiovascular Translational Research; Associate Professor of Medicine, University of Minnesota, Minneapolis

RICHARD J. HODES, Director, National Institute on Aging, Bethesda, MD

MUIN KHOURY, Director, National Office of Public Health Genomics, Centers for Disease Control and Prevention, Atlanta, GA

GABRIELA LAVEZZARI, Assistant Vice President, Scientific Affairs, PhRMA, Washington, DC

THOMAS LEHNER, Director, Office of Genomics Research Coordination, National Institute of Mental Health, Bethesda, MD

DEBRA LEONARD, Representative of the College of American Pathologists; Professor and Chair of Pathology at the University of Vermont College of Medicine; Physician Leader of Pathology and Laboratory Medicine at Fletcher Allen Health Care, University of Vermont College of Medicine, University of Vermont, Burlington

TERI A. MANOLIO (*until March 2015*), Director, Division of Genomic Medicine, National Human Genome Research Institute, Rockville, MD

ELIZABETH MANSFIELD, Deputy Office Director for Personalized Medicine, Office of In Vitro Diagnostics and Radiological Health, Center for Devices and Radiological Health, U.S. Food and Drug Administration, Silver Spring, MD

LAURA K. NISENBAUM, Research Fellow, Tailored Therapeutics, Eli Lilly and Company, Indianapolis, IN

MICHELLE A. PENNY (*until February 2015*), Senior Director, TTx Genetics and Bioinformatics, Eli Lilly and Company, Indianapolis, IN

ROBERT M. PLENGE, Vice President, Merck Research Labs; Head, Genetics and Pharmacogenomics, Merck Research Laboratories, Boston, MA

AIDAN POWER (*until July 2014*), Vice President and Head, PharmaTx Precision Medicine, Pfizer Inc., Groton, CT

VICTORIA M. PRATT, Representative of the Association for Molecular Pathology; Associate Professor of Clinical Medical and Molecular Genetics and Director, Pharmacogenomics Diagnostic Laboratory, Department of Medical and Molecular Genetics, Indiana University School of Medicine, Indianapolis

RONALD PRZYGODZKI, Associate Director for Genomic Medicine and Acting Director of Biomedical Laboratory Research and Development, Department of Veterans Affairs, Washington, DC

MARY V. RELLING, Member and Chair, Department of Pharmaceutical Sciences, St. Jude Children's Research Hospital, Memphis, TN

NADEEM SARWAR, Vice President and Global Head, Genetics and Human Biology; Chief Clinical Officer, Product Creation Headquarters, Eisai Inc., Cambridge, MA

JOAN A. SCOTT, Chief, Genetic Services Branch, Division of Services for Children with Special Health Needs, Maternal and Child Health Bureau, Rockville, MD

SAM SHEKAR, Chief Medical Officer, Health Information Technology Program, Northrop Grumman Information Systems, McLean, VA

KATHERINE JOHANSEN TABER, Director, Personalized Medicine, American Medical Association, Chicago, IL

DAVID VEENSTRA, Professor, Pharmaceutical Outcomes Research and Policy Program, Department of Pharmacy, University of Washington, Seattle

MICHAEL S. WATSON, Executive Director, American College of Medical Genetics and Genomics, Bethesda, MD

DANIEL WATTENDORF, Deputy Chief, Medical Innovations, Department of the Air Force; Program Manager, Defense Advanced Research Projects Agency/Defense Sciences Office, Arlington, VA

CATHERINE A. WICKLUND, Past President, National Society of Genetic Counselors; Director, Graduate Program in Genetic Counseling; Associate Professor, Department of Obstetrics and Gynecology, Northwestern University, Chicago, IL

ROBERT WILDIN, Chief, Genomic Healthcare Branch, National Human Genome Research Institute, Bethesda, MD

JANET K. WILLIAMS, Representative of the American Academy of Nursing; Professor of Nursing, University of Iowa, College of Nursing, Chair of Behavioral and Social Science, Iowa City

Fellow

SAMUEL G. JOHNSON (*until August 2014*), American Association of Colleges of Pharmacy/American College of Clinical Pharmacy Anniversary Fellow

IOM Staff

ADAM C. BERGER, Project Director
SARAH H. BEACHY, Associate Program Officer
MEREDITH HACKMANN, Senior Program Assistant
ANDREW M. POPE, Director, Board on Health Sciences Policy

Reviewers

This report has been reviewed in draft form by individuals chosen for their diverse perspectives and technical expertise, in accordance with procedures approved by the National Research Council's Report Review Committee. The purpose of this independent review is to provide candid and critical comments that will assist the institution in making its published report as sound as possible and to ensure that the report meets institutional standards for objectivity, evidence, and responsiveness to the study charge. The review comments and draft manuscript remain confidential to protect the integrity of the process. We wish to thank the following individuals for their review of this report:

STEPHEN LEFFLER, The University of Vermont Medical Center
TRACY A. LIEU, Kaiser Permanente Northern California
SCOTT MOSS, Epic
MARTIN PHILLIP SOLOMON, Brigham and Women's Primary Care Associates of Brookline

Although the reviewers listed above have provided many constructive comments and suggestions, they did not see the final draft of the report before its release. The review of this report was overseen by **CLYDE J. BEHNEY,** Interim Leonard D. Schaeffer Executive Officer of the Institute of Medicine. Appointed by the Institute of Medicine, he was responsible for making certain that an independent examination of this report was carried out in accordance with institutional procedures and that all review comments were carefully considered. Responsibility for the final content of this report rests entirely with the rapporteurs and the institution.

Acknowledgments

The support of the sponsors of the Institute of Medicine Roundtable on Translating Genomic-Based Research for Health was crucial to the planning and conduct of the workshop Genomics-Enabled Learning Health Care Systems: Gathering and Using Genomic Information to Improve Patient Care and Research and for the development of the workshop summary report. Federal sponsors are the Department of Veterans Affairs; Health Resources and Services Administration; National Cancer Institute; National Human Genome Research Institute; National Institute of Mental Health; National Institute of Nursing Research; and National Institute on Aging. Nonfederal sponsorship was provided by the American Academy of Nursing; American College of Medical Genetics and Genomics; American Heart Association; American Medical Association; American Society of Human Genetics; Association for Molecular Pathology; Biogen Idec; Blue Cross and Blue Shield Association; College of American Pathologists; Eli Lilly and Company; Genetic Alliance; International Society for Cardiovascular Translational Research; Janssen Research & Development, LLC; Kaiser Permanente Program Offices Community Benefit II at the East Bay Community Foundation; Merck & Co., Inc.; National Society of Genetic Counselors; Northrop Grumman Health IT; Pfizer Inc.; and PhRMA.

The Roundtable wishes to express its gratitude to the expert speakers who explored how genomic information could be gathered and used for improving patient care and research in the context of a genomics-enabled learning health care system. The Roundtable also wishes to thank the members of the planning committee for their work in developing an excellent workshop agenda. The project director would like to thank project staff who worked diligently to develop both the workshop and the resulting summary.

Contents

ABBREVIATIONS AND ACRONYMS   xix

1 INTRODUCTION AND THEMES OF THE WORKSHOP   1
Building on the Existing Learning Health Care System, 3
Putting Thoughts into Action, 6
Organization of the Workshop Summary, 7

2 ADVANCING PATIENT CARE AND RESEARCH WITH GENOMIC INFORMATION   9
The Types and Quality of Genomic Data, 10
Advancing Research and Patient Care, 14

3 TRANSLATION OF GENOMICS FOR PATIENT CARE AND RESEARCH   19
Engaging Patients, 20
Platform-Supported, Complete Learning Cycles, 24
Improving Health with a Knowledge-Based System, 28
Innovation Within Health Systems, 29
Using Genomic Data in the Clinic, 32

4 GENOMICS AND THE EHR IN A LEARNING HEALTH CARE SYSTEM   37
Leveraging EHRs for Genomics, 38
Creating a Supportive Infrastructure, 39
Managing Big Data, 43
Insurance and Regulatory Issues, 44
Data Sharing, 45

5	**REPRESENTING GENOMIC INFORMATION IN THE EHR ECOSYSTEM**	**47**
	Standards and Scale, 48	
	Use Cases, 49	
	Potential Next Steps and Considerations, 52	
6	**POSSIBLE NEXT STEPS**	**53**
	EHR Interoperability, 55	
	Clinical Decision Support, 57	
	Understanding Consumer Value and Preferences, 58	
	Closing Remarks, 60	

REFERENCES		**61**

APPENDIXES

A	Workshop Agenda	65
B	Speaker Biographical Sketches	73
C	Statement of Task	87
D	Registered Attendees	89

Boxes, Figures, and Table

BOXES

1-1 Objectives of the Workshop, 3
1-2 Clinical Care–Focused Questions to Facilitate the Workshop Discussions, 7

6-1 Possible Next Steps Proposed by Individual Workshop Participants, 54

FIGURES

1-1 The deployment of information technology within a knowledge-generating health care system can advance clinical research and care, 6

2-1 The Genomics England Clinical Interpretation Partnership (GeCIP) is intended to accelerate the adoption and implementation of research results into health care, 14

3-1 Components of the learning cycle and supportive platform, 25
3-2 The MedSeq Project is studying the use of genomic data in the care of 100 healthy patients and 100 patients with hypertrophic cardiomyopathy and dilated cardiomyopathy, 33

5-1 Interdependencies in the health care system complicate the establishment of inter-institutional project management structures, 50

TABLE

3-1 The 12 Online Genomics Educational Modules for Physicians Offered by the MedSeq Project, 34

Abbreviations and Acronyms

AAMC	Association of American Medical Colleges
CDC	Centers for Disease Control and Prevention
CLIA	Clinical Laboratory Improvement Amendments
CMS	Centers for Medicare and Medicaid Services
dbGaP	Database of Genotypes and Phenotypes
DIGITizE	Displaying and Integrating Genetic Information Through the EHR
EHR	electronic health record
ESPnet	EHR Support for Public Health
FDA	U.S. Food and Drug Administration
GeCIP	Genomics England Clinical Interpretation Partnership
HCC	hierarchical condition category
HuGENet	Human Genome Epidemiology Network
ICD	International Classification of Diseases
IOM	Institute of Medicine
MayoGC	Mayo Genome Consortia
NIH	National Institutes of Health

PCORI	Patient-Centered Outcomes Research Institute
PEER	Platform for Engaging Everyone Responsibly
ROCC	Research on Care Community
RPGEH	Research Program on Genes, Environment, and Health
SNP	single-nucleotide polymorphism

1

Introduction and Themes of the Workshop[1]

The sequencing of the human genome has facilitated a significant increase in our understanding of disease. By using individual genetic information to prevent, diagnose, and treat disease with better precision, genomics-enabled medicine promises health care that is personalized, predictive, proactive, and preventive rather than reactive, said Sam Shekar, chief medical officer at Northrop Grumman Health IT and co-chair of the workshop. On December 8, 2014, the Roundtable on Translating Genomic-Based Research for Health of the Institute of Medicine (IOM) hosted a workshop on Genomics-Enabled Learning Health Care Systems: Gathering and Using Genomic Information to Improve Patient Care and Research in Washington, DC.[2]

Technological advances have improved the accuracy of genome sequencing while decreasing the cost of obtaining it, and this is allowing for an unprecedented opportunity for practitioners to customize treatment options for patients based on their genetic signatures. As a result of this progress, huge quantities of genomic data are being generated and made available for use in the health care system. These developments—including the rapidly growing number of new technologies, the associated increase in genomes

---

[1] The planning committee's role was limited to planning the workshop. The workshop summary has been prepared by the rapporteurs as a factual account of what occurred at the workshop. Statements, recommendations, and opinions expressed are those of individual presenters and participants and are not necessarily endorsed or verified by the Institute of Medicine. They should not be construed as reflecting any group consensus.

[2] The workshop agenda, speaker biographical sketches, full statement of task, and registered attendees can be found in Appendixes A–D respectively. For more information about the workshop, see http://www.iom.edu/Activities/Research/GenomicBasedResearch/2014-DEC-08.aspx (accessed February 9, 2015).

1

sequenced, and the resulting massive amounts of new data—are enabling the ongoing transformation of the traditional, symptom-based approach to health care and treating disease into a condition-based, personalized-medicine approach, Shekar said. By integrating genomic information into the health system and delivering greater value for clinical diagnosis and treatment, precision medicine will be a disruptive force in health and health care, he said.

The integration of genomic information into clinical practice has led to discussions about how to maximize the benefits of this large amount of data. Electronic health records (EHRs), for example, could be valuable for storing and accessing clinical genomic information, as could other data sources such as self-report databases. However, the current health care system is largely unprepared to handle information on this scale (Starren et al., 2013). There is a lack of standards for such data, and it would also be useful to address such issues as interoperability, scalability, storage, privacy, security, and ethics. While individual efforts may exist to collect and use these data, it would be valuable to have a more coordinated effort that engages the broader stakeholder population in order to improve patient health and maximize the knowledge that would be obtained from integrating genomic information into health care systems.

The inclusion of genomic data in a knowledge-generating health care system[3] infrastructure is one promising way to harness the full potential of that information to provide better patient care. In such a system, clinical practice and research inform each other with the goal of improving the efficiency and effectiveness of disease prevention, diagnosis, and treatment (Ginsburg, 2014). To examine pragmatic approaches to incorporating genomics in learning health care systems, the IOM's Roundtable on Translating Genomic-Based Research for Health hosted a workshop which convened a variety of stakeholder groups, including commercial developers, health information technology professionals, clinical providers, academic researchers, patient groups, and government and health system representatives, to present their perspectives and participate in discussions on maximizing the value that can be obtained from genomic information. The workshop examined how a variety of systems are capturing and making use

---

[3] *Knowledge-generating health care system* refers to an automated system that relies upon large databases of research and patient information. Information gleaned from patients and clinical research is used in learning networks to inform clinical decisions and create a more efficient way to improve health care for future patients. This concept is also referred to as a *learning health care system* (IOM, 2012) or a *rapid learning health care system* (Etheredge, 2014).

> **BOX 1-1**
> **Objectives of the Workshop**
>
> - To explore how key pieces of genetic/genomic information can be effectively and efficiently delivered to patients and clinicians for improving care.
> - To discuss how both the health care system and genomic data can be used for evidence generation in research and in patient care.
> - To assess current best practices for using knowledge-generating/learning health care systems and which models may provide an opportunity for genomics to be used in the rapid-learning process.

of genomic data to generate knowledge for advancing health care in the 21st century. It also sought to evaluate the challenges, opportunities, and best practices for capturing or using genomic information in knowledge-generating health care systems. The objectives from the workshop are given in Box 1-1.

BUILDING ON THE EXISTING LEARNING HEALTH CARE SYSTEM

The health care system already has made great progress in building a rapid-learning system, even before the widespread incorporation of genomic information in these systems, observed Lynn Etheredge, the director of the Rapid Learning Project. Several entities, including the National Institutes of Health (NIH), the U.S. Food and Drug Administration (FDA), and the Patient-Centered Outcomes Research Institute (PCORI), have been at the forefront of this effort, and billions of dollars already have been invested in laying the foundation for such a system (see Chapter 2 for more information).

The core element of a rapid-learning health system, Etheredge said, is a computerized health system designed to use large, distributed databases and learning networks with tens of millions of privacy-protected patient records. "The goal is to learn as quickly as possible about the best medical care for each person—and to deliver it," he said.

The model that has been emerging is a common data model for data drawn largely from EHRs and claim information. This model uses distributed but networked databases, national coordinated centers, automated study designs, data quality checks, and analysis and reporting tools for computerized studies and clinical trials. The data from EHRs and claims are not necessarily as accurate as those from clinical trials, Etheredge acknowledged, but they contain valuable information that often is accurate and is representative of diagnoses from the larger population.

This system offers tremendous performance gains over what has been accomplished in the past, Etheredge said. The rapid-learning model has enabled the production of an abundance of data because many studies can be done and results can be produced much faster than before. More patients and subgroups can be studied, including seniors, children, patients with multiple diseases, minorities, and people with rare diseases. Database studies that used to take 2 years can be completed in weeks and can encompass millions of patients. For example, Etheredge said, in a study of the risk of angioedema associated with patients who take drugs to treat hypertension, 3.9 million individuals from 17 health plans that participated in the Mini-Sentinel program were examined quite quickly because the information was accessible in a database (Toh et al., 2012). Studies performed as randomized clinical trials costing millions of dollars can be completed quickly as registry trials for a fraction of the cost, as was demonstrated with the Thrombus Aspiration in ST-Elevation Myocardial Infarction (TASTE) trial in Scandinavia (Lauer and Bonds, 2014; Lauer and D'Agostino, 2013). With accessible databases, a drug safety trial that used to take months can be done in 24 hours, and knowledge that used to take decades to acquire can be generated in less than 1 year, Etheredge said.

There are other advantages to using a rapid-learning system. More questions can be answered with the data, yielding additional information that is useful to providers and patients. A larger number of researchers and learning networks can be involved, which can lead to both informal and formal collaborations. And more groups are interested in funding the research, including specialty societies, patient groups, health plans, hospital groups, accountable care organizations, and foundations. This new model, Etheredge said, has been "laying the foundation for 21st-century biomedicine as a digital science and as a system that is optimized for discovery science."

A rapid-learning system also takes advantage of the positive economics of data sharing. If each of 10 institutions shares 100 cases, then each institution gets 900 added cases in return for its sharing of 100, a return of 9 to 1. If each of 100 institutions shares 1,000 cases, the return is 99 to 1. Data sharing is a high-payoff strategy, and more data sharing multiplies benefits, Etheredge said.

Challenges to Integrating Genomics

Moving into the genomics era poses a challenge for the rapid-learning system. The current learning health care system does not integrate genomic information, and part of the challenge of such integration will be to encourage the developers of these systems to incorporate genomic data. The logical way to approach this task, Etheredge said, will be to begin to add genomic data to the common data model[4] so that genomic data can be accessed through the high-speed, high-performance research system. Key questions concerning this process are what data to add, from how many patients, and from which population cohorts. The NIH should play a significant role in answering these questions, Etheredge said, but a genomics-enabled rapid-learning health system will require collaboration among multiple government agencies and the health sector as well, including physicians, other providers, and patients.

The investments that have already been made in a pre-genomics rapid-learning health care system have created the foundations and opportunities for a genomics-enabled rapid-learning system, Etheredge concluded. Failure to act now may lead to massive amounts of genomic data being paid for by health systems but not being available for learning.

Health Information Technology Infrastructure

One of the challenges of integrating genomic information into the health care system, Shekar said, is that the massive amount of data requires a supportive information technology infrastructure for the assessment of the data. According to a 2014 report to the Agency for Healthcare Research and Quality,[5] a robust data infrastructure that can enable a learning health care system must have several features (see Figure 1-1). One is the ability to integrate various sources of information, including clinical data, genomic data, and laboratory data. The integrated data are analyzed with tools for sequence processing, managing big data analytics, and using the cloud, all with data security and safeguarded access. Findings from these analyses must be visualized in a way that makes sense and that can be communicated at the point of care and in a

---

[4]Distributed Database and Common Data Model, http://mini-sentinel.org/date_activities/distributed_db_and_data/details.aspx?ID=105 (accessed March 8, 2015).

[5]Data for Individual Health, http://healthit.ahrq.gov/sites/default/files/docs/publication/2014-jason-data-for-individual-health.pdf (accessed February 9, 2015).

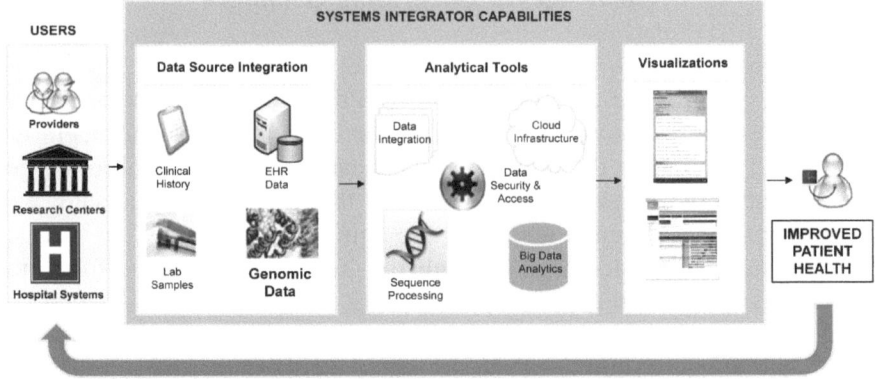

FIGURE 1-1 The deployment of information technology within a knowledge-generating health care system can advance clinical research and care.
SOURCE: Shekar, IOM workshop presentation on December 8, 2014.

brief patient encounter, Shekar said. It would be useful to measure the resulting improvements in patient health, with the results fed back into the system to improve the value and effectiveness of patient care.

PUTTING THOUGHTS INTO ACTION

A knowledge-generating health care system enabled for genomics will not be a separate system, but rather will be an extension of the current system, said Geoffrey Ginsburg, director of the Center for Applied Genomics and Precision Medicine and professor of medicine and pathology and of biomedical engineering at Duke University and co-chair of the workshop. As such, all stakeholders will be involved in shaping that system, including providers, insurers, patients, researchers, policy makers, and the health information technology community. Ginsburg offered the questions in Box 1-2 to help frame the workshop discussions about the use of genomic data within health systems.

Stakeholders will need to decide how to use the data, Ginsburg said. Health care providers will need information at the point of decision so that they are able to use it in the context of their clinical workflow, he said, and patients will need to define preferences about the use and sharing of their genomic information. All members of the health care workforce and the public will need sufficient genomic literacy to make use of new information. Researchers will need to identify and adopt best practices for

INTRODUCTION AND THEMES OF THE WORKSHOP

> **BOX 1-2**
> **Clinical Care–Focused Questions to Facilitate the Workshop Discussions (as presented by Ginsburg)**
>
> - How can health systems engage individuals to achieve health using genomic and other technologies?
> - How can systems, providers, and patients learn from failed efforts to continuously improve health and treatments?
> - How can genomic data be used to support patient-centered care?
> - How can health systems help research and care teams have access to all the data?

research using EHR-linked genomic information. The EHR vendor community will work separately and collaboratively to offer providers systems that will enable them to make more informed decisions, Ginsburg said. The health information technology community will need to design secure and interoperable genomics-enabled systems for actionable use in both health care and community settings. And policy makers will need to address the return of results, privacy, confidentiality, and education while developing regulations and economic incentives that can align all stakeholders toward the same outcomes. Health care providers will need to learn to apply genomic information to clinical decisions.

ORGANIZATION OF THE WORKSHOP SUMMARY

Following this introductory chapter, Chapter 2 considers the types and quality of genomic data to be handled by a knowledge-generating health care system. Examples are provided of pharmacogenomics data used for research as well as genetic and environmental information used to study common diseases within a learning health care system. The chapter describes efforts by U.S. government agencies and UK agencies to explore the use of large-scale genomic data to inform both research and patient care. The issue of standardizing data so that it could be re-used to maximize its potential is also addressed.

Chapter 3 considers the use of genomic information in patient care and research from different perspectives within the health care system. A foundational concept that was presented is the closed, platform-supported rapid learning cycles that provide bidirectional feedback of the analysis of data to produce results and the use of those results to change practices. It is explained that a key element of a functional rapid learning system is engaging patients and understanding their preferences for access and sharing of the data. Educating patients and physicians about the effective use of genomic data in the clinic and addressing health care disparity issues that may arise from introducing genomic information into the health care system is also discussed.

Chapter 4 examines both the potential and the challenges of integrating genomics into the EHR. Effective incorporation of the data into EHR platforms will require establishing data standards so that information is transmitted among interoperable systems and shared easily. The chapter also addresses challenges to a genomics-enabled EHR, such as determining how and what information will be shared, ensuring access to the information, establishing clinical decision support and guidance for clinicians, and improving the management of big data for enhancing clinical care.

Chapter 5 reviews the recent activities and future plans of the Displaying and Integrating Genetic Information Through the EHR (DIGITizE) Action Collaborative. The idea for this group originated with the Roundtable on Translating Genomic-Based Research for Health to engage key stakeholders in establishing a framework for data standards. The action collaborative is focusing on pharmacogenomics use cases to establish a pilot project for representing genetic test information in a structured format that can reside in the EHR ecosystem.

Finally, Chapter 6 presents possible next steps identified during workshop discussion sessions for integrating genomic data into the health care system. Several workshop participants discussed how it would be useful to make EHRs fully interoperable for genomic and other clinical information. The social science and behavioral aspect of implementing genomic information in the clinic is discussed, as is the inclusion of consumer data.

2

Advancing Patient Care and Research with Genomic Information

Important Points Highlighted by the Individual Speakers

- Integrating high-quality data into the health care system is a priority for ensuring that the best possible information is available for patient care and research. (Peterson, Risch)
- While a variety of genomic data sources exist, they are not readily accessible for use in the current electronic health record (EHR), and work to address the format of these data would create opportunities to use the information more effectively. (Risch)
- Medical information in the EHR coupled with gene sequencing information can be used as a discovery tool for identifying genetic variants associated with disease and for understanding individual response to therapeutics. (Peterson)
- Several ongoing efforts within government and the private sector are aimed at establishing data repositories for large-scale genomic information. These data could be used to demonstrate the power of rapid learning for improving patient care and informing health research. (Etheredge, Peterson)
- Data that are standardized, comparable, and consistent would facilitate the reuse of those data for discovery in multiple contexts beyond the original one. (Chute)

Ensuring the quality of the genomic data that are integrated into the health system is important to making certain that patient care is delivered and research is conducted with the best information available. Programs in the private sector, government, and at universities are currently in place or are being developed to generate genomic data that can be made accessible to local and global communities so that they can use those data to enhance patient care and impact research progress.

THE TYPES AND QUALITY OF GENOMIC DATA

There are many sources and types of genetic and genomic data, said Neil Risch, the Lamond Family Foundation Distinguished Professor in Human Genetics, the director of the Institute for Human Genetics, and a professor and former chair of the Department of Epidemiology and Biostatistics at the University of California, San Francisco. The sources of data include tests for inborn errors of metabolism (such as newborn screening tests), chromosome studies (such as cytogenetic tests), array comparative genomic hybridization, DNA-based Mendelian disorder testing, and tumor sequencing. However, a major problem is that many of the results of these tests are not easily used in research because they are often represented as PDF files in the medical record. Furthermore, while most genetic tests are reliable, the quality of the results from some of the newer tests is not universally high. For example, a study of whole genome and exome sequencing found an error rate of 0.1 to 0.6 percent, depending on the platform and depth of coverage (Wall et al., 2014). "There still needs to be more work cleaning this up before next generation sequencing is what we would consider to be clinical grade," Risch said.

If the information derived from genomic studies is to be reusable in multiple contexts, the data need to be standardized, comparable, and consistent, observed Christopher Chute, professor of medical informatics at the Mayo Clinic at the time of the workshop. For example, he said, the genomic data in the Database of Genotypes and Phenotypes (dbGaP) at the National Institutes of Health (NIH) are reasonably reusable, but the phenotypic data still lack comparability and consistency. "If we are going to generalize the research data, we need to do so in a way that we can pool [and] generate meta-analyses, reuse the data intelligently, and move on." The Mayo Clinic has what Chute termed a "local dbGaP," or the Mayo Genome Consortia (MayoGC), which pools genomic information from multiple studies across the Mayo Clinic and uses the information along with phenotype data that have been extracted from the electronic health record (EHR) for association studies.

EHRs as a Research Tool

Vanderbilt BioVU,[1] a DNA databank and biospecimens repository linked to anonymized medical records, is being used to study the associations between genes and diseases and between genes and patient responses to medications. The resource has undergone considerable growth over the past decade, said Josh Peterson, an assistant professor of biomedical informatics and medicine at the Vanderbilt University School of Medicine. It now contains close to 200,000 samples, about 170,000 of which are adult and the remainder pediatric. About 90,000 have been genotyped with a high-density platform, usually a genome-wide association study or exome chip. Studying biobank data and corresponding phenotype data in EHRs can confirm known genetic associations and therefore be used as a discovery tool in genomics (Ritchie et al., 2010).

However, before BioVU is used as a discovery tool, the method needs to be validated, Peterson said. For example, in a study to predict cardiac events using genetic variants in patients receiving clopidogrel, 260 of 591 phenotyping cases were confirmed as "definite cases," or patients who were prescribed clopidogrel following a myocardial infarction or percutaneous coronary and who then experienced one or more recurrent cardiac events (Delaney et al., 2012). Once the high-quality data were generated, an analysis of them demonstrated that adverse recurrent coronary events were correlated with *CYP2C19* and *ABCB1* but not with *PON1* in that patients with specific variants of the first two genes were more likely to experience those events than control patients who did not have those variants.

The BioVU resource has also been used to link phenotype data with genomic data, which Peterson referred to as PheWAS data. For example, in an association study of single-nucleotide polymorphisms (SNPs) and EHR-derived phenotypes, *IRF4*, which was known to be linked with hair and eye color, was newly associated with actinic keratosis, a skin condition that may progress to cancer (Denny et al., 2013).

The Pharmacogenomic Resource for Enhanced Decisions in Care and Treatment program at Vanderbilt focuses on germline pharmacogenomic variants and has genotyped about 14,000 patients. Selected pharmacogenomic data are reported to the EHR so that providers receive clinical decision support that takes into account genomic variants. The quality of the genotyping data is very high, Peterson reported, including

---

[1] Vanderbilt Research, https://victr.vanderbilt.edu/pub/biovu/index.html?sid=194 (accessed March 4, 2015).

nearly 100 percent call rates for actionable variants and 100 percent concordance on repeat samples. This is important, he said, because low-quality data that can be obtained for trouble spots such as the highly polymorphic *CYP2D6* loci can also be a problem for rapid learning health systems over time because of low replication accuracy.

The Electronic Medical Records and Genomics Network (Friedman et al., 2015) has also demonstrated that it is possible to use EHRs to do genomic research. Cohorts could be generated across multiple medical centers with shared algorithms in a reproducible and consistent way. The resulting studies have not been perfect, Chute said, "but they are clearly demonstrating that you can consistently and collaboratively leverage disparate and heterogeneous health records in a way that you can use that information for underlying research." Bielinski et al. (2014) showed that MayoGC can be used successfully as a research tool to study genetic variants associated with bilirubin levels using data from individuals enrolled in three NIH-funded studies at the Mayo Clinic.

An area where BioVU has been particularly helpful for clinical implementation has been in creating warfarin dosing algorithms. Two commonly used algorithms are from the International Warfarin Pharmacogenomics Consortium and WarfarinDosing.org. The current difficulties in using genetic data to determine warfarin dosing may arise from the fact that errors in the algorithms are still too large and need to be reduced through further research, Peterson said. A new algorithm was developed at Vanderbilt in response to disparate results from studies in which warfarin dosing was guided by genetics, he said (Ray, 2013). Adverse events are tracked, but one limitation, Peterson acknowledged, is that re-contacting patients is not an option.

Genomic data are often de-identified prematurely, Chute said, but if the data are to maintain maximum usefulness, linkages need to be maintained between the clinical phenotypic data and the underlying genomic data. "I'm all for privacy and security," he said, "but the importance of maintaining the consistency and linkage of the clinical information cannot be underestimated."

The 100,000 Genomes Project

In the United Kingdom the 100,000 Genomes Project[2] is intended to establish a genomic program that is transparent to patients, that will sup-

---

[2] The 100,000 Genomes Project, http://www.genomicsengland.co.uk/the-100000-genomes-project (accessed March 4, 2015).

port scientific and medical inquiry, and that will foster the development of an industry in genomics within the United Kingdom. By the end of 2017, the sequencing of 100,000 genomes will be complete, said Tom Fowler, the director of public health at Genomics England.

The genome sequences will be generated from National Health Service patients with rare inherited diseases, cancers, and pathogens. This specific focus was chosen, explained Fowler, because those were the three areas deemed most likely to result in expeditious translation from genomics research to practice. A key feature of the 100,000 Genomes Project is learning to use genomic technology and data in the health care system. For example, there is interest in "deep phenotyping" patients, or providing comprehensive detail about the components of patients' phenotypes, because this may lead to improved diagnoses for diseases (Robinson, 2012).

The creation by the 100,000 Genomes Project of National Health Service centers for genomic medicine will result in samples and data being provided to the broader collaborative. Created at various institutions around the country, these centers are investing internal resources in this project, Fowler said. They also present an opportunity to move toward a hybrid approach to clinical care and research, in which both clinical care and research happen at once rather than being separate enterprises.

In addition to the genomic medicine centers, the 100,000 Genomes Project has created the Genomics England Clinical Interpretation Partnership (GeCIP), which is a mechanism for bringing the National Health Service and academic communities together to use the data that have been collected in order to analyze and assess how the genome dataset could be interpreted for clinical use. By opening up databases developed by individual researchers, the partnership will be able to take advantage of the capabilities of an entire community, including clinicians, and academic researchers, Fowler said. All generated data are contributed to the Genomics England Dataset and are available to all, with the intellectual property owned by Genomics England but freely licensed. The goal is to greatly accelerate the use of research-based results in health care (see Figure 2-1).

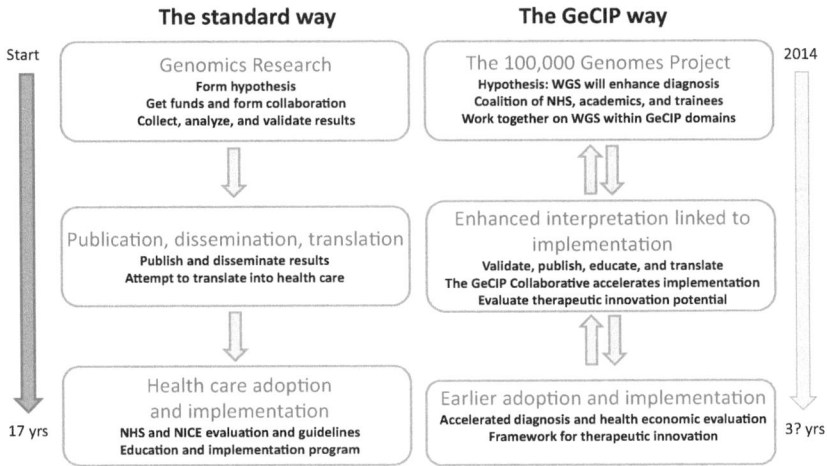

FIGURE 2-1 The Genomics England Clinical Interpretation Partnership (GeCIP) is intended to accelerate the adoption and implementation of research results into health care.
NOTE: GeCIP, Genomics England Clinical Interpretation Partnership; NHS, National Health Service; NICE, National Institute for Health and Care Excellence; WGS, whole genome sequencing.
SOURCE: Fowler, IOM workshop presentation on December 8, 2014.

ADVANCING RESEARCH AND PATIENT CARE

Several U.S. government initiatives are exploring ways to use genetic and genomic data to further research. Etheredge reported that the National Cancer Institute is developing and testing a new master protocol trial system[3] in 200 collaborating centers which could become the basis for a much faster trial system for genetically informed research (Ledford, 2013). Genetic profiling is being used initially to determine which of five different treatment modalities will benefit a given patient the most so that the patient can be assigned to the most promising of five parallel treatment arms. The result of such assignments could be reduced cost and faster and smaller trials, since the cohorts can be organized genetically.

---

[3]Lung-MAP launches: First precision medicine trial from National Clinical Trials Network, http://www.cancer.gov/newscenter/newsfromnci/2014/LungMAPlaunch (accessed February 18, 2015).

Patient groups are enthusiastic, because the people in trials can get the best therapies available based on predictive models, and this type of approach could be greatly expanded, Etheredge said.

NIH is also working on a conceptual framework for what it calls The Commons,[4] a cloud-based platform in which databases from publicly supported studies are shared among the biomedical research community—grantees, applicants, government agencies, the private sector, and others. Developing such a computing infrastructure would allow for the sharing of existing data in an accessible manner in order to foster the development of new ideas and knowledge by reusing data and avoiding duplication of studies. Grants include funds for curating and archiving databases and "vouchers" to allow researchers to access, analyze, and use the data, Etheredge said. The standards and data developed through the centers of excellence under the Big Data to Knowledge[5] initiative would provide information that could be piloted as part of the emerging Commons. Some of the centers of excellence have a specific focus on genomics, and they are working to build an interoperable infrastructure. This would allow clinicians and researchers to share large-scale genomic data and to mine the information with computational engines that would inform research and, eventually, patient care.

There are other opportunities for large-scale data to be used in rapid learning systems. The Centers for Disease Control and Prevention (CDC) is expanding and enriching a genomics-enabled research system for epidemiology and public health science—for example, through its Human Genome Epidemiology Network, the HuGENet initiative. The goal of the program is to "translate genetic research findings into opportunities for preventive medicine and public health."[6] In collaboration with the Harvard Pilgrim Health Care Institute and Children's Hospital, CDC has also developed a real-time national tracking and rapid learning network for public health emergencies, called EHR Support for Public Health, or ESPnet.[7] Using data from initiatives such as these in learning systems could provide insights into how information could be used in preventive medicine and improving public health.

---

[4] The Commons, https://pebourne.wordpress.com/2014/10/07/the-commons/#_ftn2 (accessed March 12, 2015).

[5] Big Data to Knowledge (BD2K), http://bd2k.nih.gov/#sthash.hw61mFU4.dpbs (accessed March 13, 2015).

[6] Human Genome Epidemiology Network, http://www.cdc.gov/Genomics/hugenet/default.htm (accessed February 18, 2015).

[7] ESPnet, http://esphealth.org/ESPnet/images/overview.html (accessed February 18, 2015).

Other Genetic Research Resources

Kaiser Permanente started the Research Program on Genes, Environment, and Health (RPGEH), of which Risch is a lead co-investigator, to "examine the genetic and environmental factors that influence common diseases such as heart disease, cancer, diabetes, high blood pressure, Alzheimer's disease, asthma, and many others."[8] The program uses Kaiser Permanente's comprehensive EHR, supplemented with behavioral and demographic data from surveys, information on environmental exposures, and collected biospecimens, to study common diseases. To date, Kaiser Permanente has gathered about 200,000 saliva and blood specimens, along with survey data on demographics, health history, family history, smoking, alcohol use, diet, physical activity, and reproductive history. Although Kaiser Permanente is largely a clinical enterprise, it has invested in research, and the combination ultimately will translate into benefits for patients, Risch said.

RPGEH is intended to advance research by creating a large databank of genetic and other medical information along with lifestyle, demographic and environmental data that will be accessible to the Kaiser Permanente Division of Research and to collaborating scientists from other institutions. The long-term goal is to identify the genetic and environmental basis for common age-related diseases along with factors that influence healthy aging and longevity. The specific aims of the program, Risch said, are to conduct genome-wide genotyping of more than 675,000 markers on 100,000 participants in RPGEH; to assay telomere lengths for the same 100,000 samples; to develop customized genome-wide SNP arrays and use these arrays for genotyping; to merge, with patient consent, the genomic and telomere data with the EHR, survey, and environmental data in a research database; and to provide collaborative access to the data.

The group of subjects participating in RPGEH is 58 percent female and has an average age of about 65, said Risch. It is 78 percent white, 11 percent Latino, 8 percent Asian, and 3.5 percent African American, with more than half the participants having been members of Kaiser Permanente for 20 or more years. Comprehensive electronic records go back to 1995, with the physician notes being accessible from 2006. Information on cardiovascular disease, psychiatric disorders, cancer, diabetes, and other conditions is available for many thousands of people, along with

---

[8]The Research Program on Genes, Environment, and Health, http://www.dor.kaiser.org/external/DORExternal/rpgeh/index.aspx (accessed March 13, 2015).

data from electrocardiograms, magnetic resonance imaging, computerized tomography scans, mammographies, ophthalmologic exams, lipid panels, other serum chemistries, blood pressures, body mass indexes, and other health measures.

The genotyping was completed at the Institute for Human Genetics at the University of California, San Francisco, and it produced very high quality data, Risch said. Genome-wide association studies have led to the identification of more than 600 contributing genetic variants—approximately one-third of which were novel—which are associated with a variety of traits and diseases extracted from EHRs, ranging from blood pressure, cholesterol levels, and QT intervals to prostate cancer and diabetes. Data can be accessed in two ways: through a Web portal at Kaiser Permanente, where a committee reviews applications for the use of datasets by qualified researchers, and through dbGaP.[9] In 2014, Kaiser Permanente made a large deposit of data into dbGaP—from 78,000 people who participated in the Genetic Epidemiology Research on Adult Health and Aging project, part of the RPGEH.[10] The genetic data are housed in a separate database from the EHR data and are currently available only for research purposes.

Opportunities

Existing programs that generate data and foster research should be examined so that upgrades over the next couple of years could be planned to determine how they could facilitate rapid learning, Etheredge said. At FDA, the opportunities to leverage current programs into rapid learning systems that incorporate genomics could include national registries, standardized data, and coverage with evidence development initiatives, he said. The Sentinel[11] system is accruing data on more than 50 million patients annually and has 380 million patient-years in its database. It could be extended into a tracking and registry system for effectiveness as well as safety. Clinical and scientific databases are being made publicly available, and oversight of predictive models could inform

---

[9]Resource for Genetic Epidemiology Research on Adult Health and Aging (GERA), http://www.ncbi.nlm.nih.gov/projects/gap/cgi-bin/study.cgi?study_id=phs000674.v1.p1 (accessed March 4, 2015).

[10]Kaiser, UCSF dump data from large genomic study into dbGaP, https://www.genomeweb.com/informatics/kaiser-ucsf-dump-data-large-genomic-study-dbgap (accessed March 13, 2015).

[11]FDA's Sentinel Initiative, http://www.fda.gov/Safety/FDAsSentinelInitiative/ucm 2007250.htm (accessed February 18, 2015).

the public about benefits and risks beyond patient package inserts.

PCORI is expanding its PCORnet capabilities in collaboration with NIH, FDA, and other agencies, Etheredge said. Potential upgrades could include identifying patient-centered research needs for genomics-enabled health care, with national work plans for who is accountable for answers to priority questions and by what time. PCORI also could engage patient groups, professional societies, health plans, hospital groups, accountable care organizations, and others for the collaborative funding of comparative effectiveness research using fast, affordable rapid learning systems. It could develop predictive models for patients and physicians to compare the benefits and risks of various options. The Department of Veterans Affairs also has plans to employ a learning health care approach with veterans who are diagnosed with non-small-cell lung cancer (Ray, 2015). The results from gene sequencing panels will be used to direct therapy, and the information will also be used for research purposes.

The Centers for Medicare & Medicaid Services (CMS) could support a genetics-enabled rapid learning center system for Medicare and Medicaid, Etheredge said. All cancer data in the systems could be collected and reported to a national privacy-protected cloud system, with coverage for genetic sequencing and analysis and predictive services. The CMS Innovation Center[12] could test and advance best practices in genomics-enabled cancer care, using pay-for-performance to improve quality. Working with FDA, the center could use coverage with evidence development to support genomics-enabled medicine, such as with new cancer treatments, and it could collaborate with the American Society of Clinical Oncology on a rapid learning cancer system, Etheredge said.

---

[12]The CMS Innovation Center, http://innovation.cms.gov (accessed February 18, 2015).

3

Translation of Genomics for Patient Care and Research

Important Points Highlighted by the Individual Speakers

- Patients are very involved in their own health care and are producing their own health-related data, including genomic data. Understanding preferences for data use and communicating effectively in a fair and transparent way with the public about how information is used will be key to engaging the larger population in sharing their data for research. (Baker)
- Patient trust can be earned and maintained through good data practices, including establishing confidentiality policies, data encryption, and multifactor authentication. (Chute)
- Deriving clear standard consent language could reduce the burden on institutions, which today largely develop their own consenting mechanisms, and could provide transparent information for patients about the use of their data. (Baker, Chute, Fowler, Moss)
- A health care system in which an infrastructure supports complete learning cycles that encompass both the analysis of data to produce results and the use of those results to develop changes in clinical practices is a system that will allow for optimal learning. (Friedman)
- Using genomic data could improve population health and contribute to solving many care management problems. Starting with areas that can lead to a return-on-investment may encourage leaders of health systems to engage in these efforts. (Hill)
- Just-in-time information, guidelines for clinical action, and more information on the clinical utility of genetic testing would help physicians make effective use of genomic information and integrate it in their practices similarly to other medical test information. (Vassy)

Efforts are ongoing to facilitate the incorporation of genomics into health systems by understanding patient preferences, educating physicians, and addressing health care disparities. Designing a closed, platform-supported system could help enhance the flow of genomic information through a learning health system, increasing efficiency and bringing added value to patient care and health.

ENGAGING PATIENTS

Patients are consumers of health products and services, said Dixie Baker, a senior partner at Martin, Blanck & Associates, and they are also the "primary source of the information and data that we need to create this learning health system."

Today's consumers are much more involved in their own care and in the health of their families than consumers in the past, Baker said. In January 2014, for example, nearly one-third of all U.S. smartphone owners (46 million unique people) used fitness and health apps.[1] It has been predicted that by 2017, 30 percent of U.S. consumers will be wearing a device to track food consumption, exercise, heart rate, and other critical vital signs.[2] More than 500,000 consumers have directly purchased DNA testing services, with no evidence of psychological harm and some evidence of positive behavior changes.[3] Consumers willingly contribute their data and biological samples to medical research—when their permission is sought. But if the data are being used for research without permission, dramatic pushback can occur.

As computational power and the amounts of clinical and genomic data continue to grow, medical knowledge and the quality of health care will continue to increase. However, risks to personal privacy will also grow, creating the possibility that consumers will limit the amount and quality of the data that they make available to health care providers and to researchers. A high percentage of consumers are concerned about the

---

[1]Hacking health: How consumers use smartphones and wearable tech to track their health, http://www.nielsen.com/us/en/insights/news/2014/hacking-health-how-consumers-use-smartphones-and-wearable-tech-to-track-their-health.html (accessed March 4, 2015).

[2]Will an app a day keep the doctor away? The coming health revolution, http://www.forbes.com/sites/ciocentral/2013/09/08/will-an-app-a-day-keep-the-doctor-away-the-coming-health-revolution (accessed March 4, 2015).

[3]Regulation: The FDA is overcautious on consumer genomics, http://www.nature.com/news/regulation-the-fda-is-overcautious-on-consumer-genomics-1.14527 (accessed March 4, 2015).

privacy and security of their medical information, Baker said. Many U.S. consumers with chronic conditions want to control their health information, but roughly half believe that they currently have very little control over that information.[4] DNA is inherently unique to the individual, rendering it the ideal "biometric identifier"—1 of the 18 data elements of identifiability defined by the Health Insurance Portability and Accountability Act. Even without a name or phenotype linkage, DNA includes many clues for narrowing the identity possibilities—and it can be obtained from objects as ubiquitous as discarded coffee cups. Access to an individual's DNA also poses a substantial privacy risk for blood relatives, who most likely did not consent to access.

Strong security protection, fair information practices, and "no surprises" will enable the learning health system to emerge, Baker suggested. Consumers generally do not want complex consent forms when they become involved in biomedical research. Rather, they want adherence to principles that have been adopted around the world to gain their trust, including transparency, being asked permission, having access to their own health information, and being provided with knowledge of how their information will be used.

The Platform for Engaging Everyone Responsibly

The nexus of issues associated with patient consent, patient privacy, and data access offers a variety of challenges to integrating genomics into the learning health care system. Patients need to have a way to indicate their preferences concerning their genomic information, said Scott Moss, a research informatics software developer with Epic. These issues are less technical. In the area of consent, for example, "there needs to be some consistent direction," Moss said, possibly by creating a toolbox of best practices.

A new platform has been developed to address issues of sharing health information in a secure manner that takes into account individual preferences for data sharing and at the same time provides ease in accessibility of the data by the medical and research communities, Baker said. By using consent-management tools developed by Private Access, Genetic Alliance has developed a participant-centric research platform

---

[4]Accenture's 2014 Patient Engagement Survey, http://www.accenture.com/us-en/Pages/insight-trends-life-sciences-consumer-perceptions-about-emr-benefits.aspx (accessed April 16, 2015).

known as the Platform for Engaging Everyone Responsibly[5] (PEER) which enables individuals to make their health information available to researchers. It has three components: a data entry element, a privacy layer, and a data query module. The data entry component is accessed through a customizable consumer portal that can be embedded in any webpage. The entry allows access to de-identified health data and personal contact information. The data query component can be accessed through a researcher portal, which today takes place largely through a tool called RecruitSource (Terry et al., 2013). Searches, alerts, and access requests are handled in this section according to individually defined permissions. In between the data entry and the query components, a privacy layer serves as a filter to create and manage permissions for sharing patient data wherever the data reside and at any level of granularity. Participants establish their own sharing preferences based on a simple "stop light" metaphor: allow, deny, or ask me. PEER enables the expression of the full spectrum of personal views about privacy and sharing of health information, with the ability to adjust settings dynamically as one's values and priorities change over time, Baker said. Individuals can define who can discover their information, who can download and use their information, and who can contact them directly if the situation makes it necessary to do so.

Genetic Alliance has created a virtual guide that people can use to establish their preferences. People can choose conservative, moderate, or liberal options, or they can go through and select each setting on their own. Among advocacy communities, about 85 percent of individuals release their data for all purposes, about 10 percent say "ask me," and 5 percent decline to release their data, said Sharon Terry, the president and chief executive officer of Genetic Alliance.

Baker said that the PEER system is designed to increase the number of people involved in clinical trials. Customizable PEER entry points are easily embedded into any website, and a smartphone app provides mobile access. The system also makes it possible to engage people beyond the groups established for families with genetic diseases.

Alignment is a critical issue, not only between researchers and clinicians but among those groups and consumers, Baker said. Consumers are interested and eager to be involved in the overall ecosystem, but they face barriers to becoming involved. "We need to take that challenge on," she said.

---

[5]Platform for Engaging Everyone Responsibly, http://www.geneticalliance.org/programs/biotrust/peer (accessed February 24, 2015).

Patient Trust, Consent, and Opting In

Establishing and maintaining patient trust is essential to demonstrating that patient information can be used securely and responsibly, Chute said, adding that the medical genomics community should "engage the public and our patients to trust that we are using genomic information respectfully and productively to enhance our understanding and discovery, and then establish an atmosphere of confidentiality," he said. The way to maintain trust of patients and the community, he continued, is through consistent data practices, including establishing and enforcing confidentiality policies, maintaining encryption, separating identifiers after dataset linkage, prohibiting any clinical data on portable devices, and using multifactor authentication.

A specific barrier to genomic research is patient consent, said Chute and Gail Jarvik, who holds the Arno G. Motulsky Endowed Chair in Medicine and heads the Division of Medical Genetics and the Northwest Institute of Genetic Medicine at the University of Washington School of Medicine. The development of standard consent language is an obvious need, Chute said. Today, every academic research center and test development company generates its own consenting mechanisms. There is no reason why every academic medical center should have to reinvent clear language because this language could be widely shared, he said. In addition, communication among and between clinicians and families can be poor, yet these routes of communication are critical in conveying and understanding genomic information. Only a few percent of patients will have incidental findings from a genomic test, Jarvik said, yet they all need to be consented for incidental findings, and this takes time. Patients also need to be able to opt out of genomic data linkages, Chute added.

Many of the policy issues facing the 100,000 Genomes Project are similar to the policy issues related to the use of genomic information in the United States, Fowler said. Patients are granting consent to link to lifetime health records, with the information being pseudo-anonymized in a research dataset. However, the consent forms are overly complex, Fowler said, and they will need to be simplified as the project progresses. To further protect the privacy of patients, researchers have to use the data within a specific infrastructure.

So far, the greatest limitation on the conduct of research has been the opt-in system for patient involvement, said Risch. When the use of genetic data is considered standard care, and when genomic sequencing is performed as routinely as other clinical tests are (results from which are broadly available for research), the need for opt-in may change, Risch observed. Aside from limited Mendelian carrier screening and cytogenetic studies, the use of genetic data in clinical research has required participant consent. Another current limitation for research, he said, is that data-analysis-only research proposals are not well received at the National Institutes of Health. However, ethical concerns are likely to still engender discussion, for example, regarding the return of results and their implications, the confidentiality and security of data, and the creation of genetic data for research purposes without current clinical relevance.

One way to address some of privacy concerns is to sort genetic information into categories that have different implications, Risch said. For instance, the information could be considered predictive when an individual already has symptoms and the purpose of genetic testing is to determine a diagnosis. When an individual has a family history of a disease but does not have the disease, testing may be indicated, and the findings could be considered incidental. Kaiser is exploring the use of this type of structure for genetic information and how it might be used in its system, Risch said. One thing that would help the clinical and research enterprises work together would be standards for representing clinical information in translational and discovery research. These standards, Chute said, should be aligned with, if not derived from, clinical standards, which would promote the secondary use of data for research.

Another issue involves not only patient access to results, which is governed by a number of regulations, but the responsibility of laboratories to maintain results over time. Will data be kept for a patient's lifetime? If so, how will new interpretations of the data be conveyed to patients? "This is an area that needs a lot more exploration," Peterson said.

PLATFORM-SUPPORTED, COMPLETE LEARNING CYCLES

In a health care system that can learn, every patient's characteristics and experiences are available for study, best practice knowledge is immediately available to support decisions, and improvement is continuous and routine, said Charles Friedman, the Josiah Macy Jr. Professor at the University of Michigan Medical School. Such a system requires complete learning cycles, each of which consists of two halves, he said (see Figure 3-1). The first half includes the assembly and analysis of data and the interpretation of results once a decision is made to study a problem of

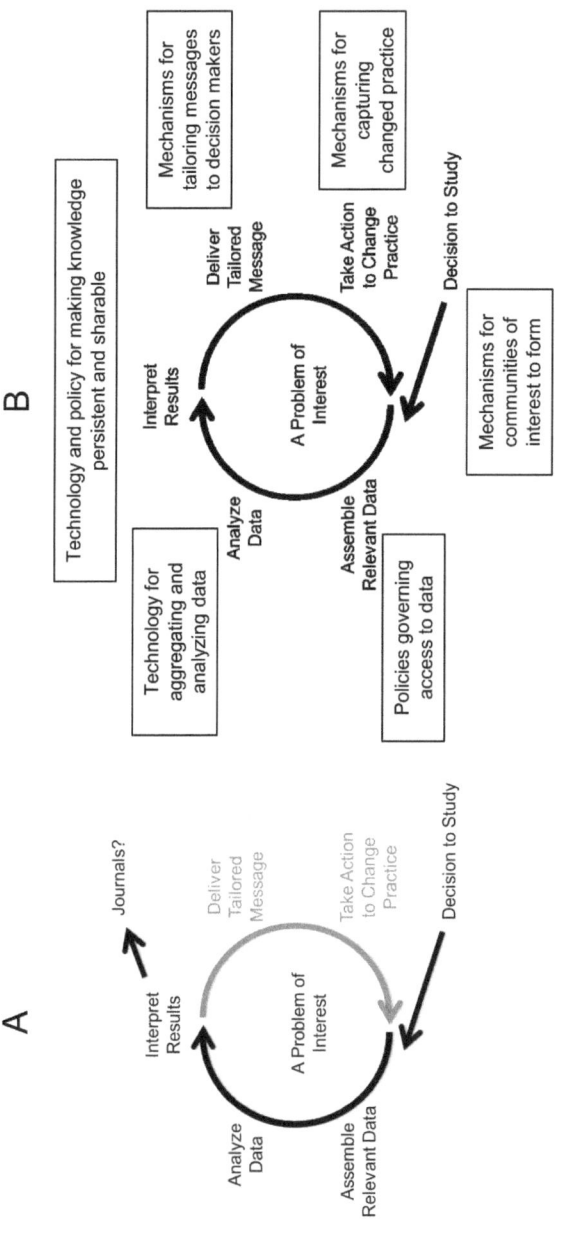

FIGURE 3-1 Components of the learning cycle and supportive platform: (A) an incomplete cycle (black arrows), and (B) a complete cycle with supportive infrastructure.
SOURCE: Friedman, IOM workshop presentation on December 8, 2014.

interest. The second half of the cycle includes the use of results to develop and deliver tailored messages, which in turn lead to action to change practice. This complete learning cycle then drives subsequent iterations of learning (Friedman et al., 2015). Currently, many projects that are establishing the basis for learning systems generally support only the first half of the learning cycle, not the second half.

Today, this learning cycle is often an open loop, Friedman said. Once the first half is completed, results are sent to journals to be published, after which it can take years to translate the results into practice (see Figure 3-1). "There seems to be a consistent belief that if we get the data and analytics side of this right, everything else is going to fall into place," he said. "But everybody knows this isn't true. In fact, maybe the harder part of the problem, as challenging as the data and analytics are, is the feedback side, where we're engaged in needing to change human behavior." The decision support systems now in place are a primitive and largely ineffectual version of what needs to be done to drive change and improvement through complete learning cycles, Friedman said.

Simultaneous Learning Cycles

To create health care systems that can learn, delivery systems and research networks must run many complete learning cycles simultaneously, Friedman said. This requires an infrastructure that makes learning effective, sustainable, and routine, with the accompanying economies of scale.

Friedman laid out the technology and policy components of a platform for such an infrastructure (see Figure 3-1). This sort of platform would have shareable and interchangeable components, thus distributing costs and making it possible for the platform to be built once and used repeatedly. Some learning cycles may cycle more slowly (e.g., clinical trials), while others would cycle faster (e.g., public health concerns such as disease outbreaks). "Without a platform, each learning cycle will develop its own—probably suboptimal—methods for learning, and there will be no economy of scale," he said. "The cost of setting up the thirteenth platform will be equal to the cost of setting up the first."

The development of the full platform could lead to a national or even a global learning system. As Friedman noted, the Internet grew out of a similar situation. A small kernel of standards was developed. Once those common standards were developed, everyone shared those standards while innovation continued to flourish around them. "We need something for the learning health system that is analogous to that small kernel of common standards to bind everything together, that will allow the kind of innovation at the edges that everyone wants to see," Friedman said.

Only if knowledge is persistent can it be improved and shared effectively through this platform, Friedman said. It will also be necessary to have mechanisms in place to tailor messages to decision makers and to capture changing practices. "Everyone's interests here are in alignment," he said. "Everybody wants the same thing. The challenge is how to make these aligned interests into a set of activities that will bring the benefits that are commonly desired."

Health Care Disparities

Groups adversely affected by disparities could be among those that benefit most from a learning health care system, observed both Baker and Jason Vassy, a primary care physician and clinician–investigator at Harvard Medical School, the Veterans Affairs Boston Healthcare System, and Brigham and Women's Hospital. For example, patients who use emergency departments often tend to be from lower socioeconomic statuses, and more data might be in the system for them than for other people. "Depending on what clinical dataset you're talking about, you might be able to improve some disparities," Vassy said.

Alexander Ommaya, the senior director of implementation research and policy at the Association of American Medical Colleges (AAMC), brought up the importance of measuring and tracking inequalities. Health care systems can evaluate the impact of their interventions on disparities, he said. "To assume that, just because we provide care and [that] we're providing it to everyone equally, we're going to solve this problem is naïve." Not everyone has access to the health care system. "We need to focus on it and develop specific interventions to address it." Friedman observed that the problems that get attention are those around which communities of interest form and generate enthusiasm for solutions. If communities of interest form around reducing disparities, then learning cycles could take shape around those issues. "Let's look at ourselves and decide what's important," Friedman said.

IMPROVING HEALTH WITH A KNOWLEDGE-BASED SYSTEM

More than half the drugs used in the United States do not work in the patients for whom they are prescribed, said Colin Hill, the chief executive officer and a co-founder of GNS Healthcare. "This, for me, points to why we need a learning health care system," he said. "It's not good enough to drive predictive models and diagnostics with data from electronic medical records and claims." Genomics-related data will help clinicians understand a patient's genetic variations and how they affect the response to therapy.

GNS Healthcare is a company that creates analytic solutions for improving population health in a cost-effective manner. These solutions are delivered as cloud-based software to health plans, health care providers, and pharmaceutical companies. The software handles data from EHRs, pharmacy and medical claims, patient registries, and other sources through machine learning platforms in order to determine what treatments will work for individual patients. As Hill put it, his company reverse-engineers causal networks so that patient characteristics can be combined with potential treatments to predict clinical and economic outcomes.

The company has applied this process to medication non-adherence, metabolic syndrome, preterm birth, and a variety of other health problems. For example, in a project with Aetna, GNS Healthcare developed a predictive model to identify which people are most likely to develop metabolic syndrome in the next year, along with the risk factors that contribute most to that development (Steinberg et al., 2014). Health care resources can then be targeted to those individuals where they can make the greatest impact. In addition, with heterogeneous data collected largely for other purposes, it was demonstrated that decreasing an individual's waist circumference and improving an individual's blood glucose levels produce the largest benefits for subsequent risk and medical costs for metabolic syndromes.

Genomic data could greatly extend the benefits of this approach, Hill said. For example, he described a project with Inova Translational Medicine Institute that combines sequence data with EHRs and other data types to predict which women are most likely to have a preterm birth. The goal is to identify the underlying causal mechanisms of preterm birth, predict personalized preterm birth risk, and accelerate the discovery of new diagnostic tools and treatments.

Gathering and analyzing more data from various populations could solve many care management problems, Hill predicted. What will ultimately make this approach feasible will be the return on investment that it produces.

INNOVATION WITHIN HEALTH SYSTEMS

Building a learning health care system requires that research, educational, and health systems be coordinated, said Ommaya. But today the linkages among those need a solid infrastructure to support a learning health care system, he said.

As an example of trying to help build this infrastructure, Ommaya described the AAMC's Research on Care Community (ROCC), which consists of researchers and clinical providers engaged in implementation research to improve the quality, safety, health equity, and outcomes of their patient populations. Since 2012, membership has more than doubled to roughly 250 people representing about 140 institutions.[1] According to a survey performed shortly before the workshop, about 90 percent of ROCC members who conduct research are currently engaged in collaborative research projects with clinical colleagues, and about two-thirds are currently engaged in collaborative research projects with teaching faculty, Ommaya said.

ROCC also offers Learning Health System Champion and Pioneer Research Awards[2] to encourage collaboration among researchers, educators, and health systems. Champion awards of $5,000 recognize best practices in research, health system, and education collaboration. Pioneer Awards of $10,000 support the implementation of systematic changes to enhance research. As an example, Ommaya discussed the Clinical Research Database at the Loyola University–Chicago Stritch School of Medicine, which is a large-scale, easy-to-use de-identified clinical data structure that provides population health information to everyone with EHR access. There is also an option to make a connection to the institutional review board submission system for access to identified data. Access to the database has alleviated the bottleneck that previously occurred when, prior to implementing the research database, requests were submitted to obtain data from the information technology department. The database features a Web application for casual users, such as faculty

---

[1]Research on Care Community, https://www.aamc.org/initiatives/rocc/about (accessed April 16, 2015).
[2]AAMC Learning Health System Champion and Pioneer Research Awards, https://www.aamc.org/initiatives/rocc/funding (accessed March 4, 2015).

members and in-house staff, and a number of tools for advanced users, such as analysts and the bioinformatics staff. The system contains data on 2 million patients and on more than 7 million patient encounters and has "really enhanced research activity at Loyola," Ommaya said.

As another example, Ommaya mentioned the Community Engagement Studio, which provides an opportunity for researchers to recruit participants or "stakeholder experts" who represent the population of interest for a study. It conducts structured sessions that engage patients, consumers, and other non-academic stakeholders appropriate to the study needs including study design, recruitment, dissemination, implementation, and consent. The Community Engagement Studio also helps prepare participants to understand what research is and what their role in a given study will be. The program also offers training for researchers on how to engage with participants successfully.

The Learning Health System Champion and Pioneer Research Awards foster collaborations and the "type of capability that you need for the learning health system," Ommaya said. This is a model that AAMC is evaluating for building capacity among their member institutions, particularly between the medical schools and the teaching hospitals.

Managing Priorities in a Health System

Health systems are being inundated with new forms of data—not just from genomics, but from gene expression profiles, proteomics, and other high-throughput technologies, said Fred Sanfilippo, the director of the Healthcare Innovation Program at Emory University and the Georgia Institute of Technology and a professor of pathology and laboratory medicine at the Emory University School of Medicine. Furthermore, many other factors besides a person's genetics affect health outcomes, including behaviors, social circumstances, and environmental exposures.

A number of challenging issues surround the organization and function of academic health centers, such as how to establish priorities and allocate resources. For these institutions to become learning health systems, research, education, and clinical care need to be aligned, Sanfilippo said. Today, that is often not the case. The hospital system, faculty members, and universities typically have different missions, emphases, priorities, and values. This observation applies to other parts of these systems as well, including the financial, planning, communications, information technology, human resources, and community outreach components. As Friedman noted earlier in the session, researchers have generally been given incentives to publish and not necessarily to establish connections with health systems. The leaders of health systems and the research enterprise are now thinking about ways to generate value from each other, which has created opportunities to bring these communities together.

One possible way to address the organizational challenges is to have a single leader oversee the enterprise, but that does not necessarily result in alignment, Sanfilippo said. A recent study of 84 university health systems found no correlation between the structural organization of the academic health center and its performance, either in the research, clinical, educational, or financial areas (Keroack et al., 2011). However, functional alignment, as measured by 12 different parameters, did correlate with those outcomes, with capital planning, strategic planning, and communication showing the highest correlation (the others were budgeting, financial reporting, program planning, chair hiring and firing, chair evaluation, medical directors, financial transfers, business development, and information systems).

More broadly, the culture of an institution can change outcomes, Sanfilippo said, echoing a point made earlier by Friedman. For example, when The Ohio State University Medical Center moved from a passive culture that lessened success to a constructive culture in which innovation and teamwork are encouraged, the academic, clinical, and financial performance improved dramatically (Sanfilippo et al., 2008). Each institution has a different culture, and changing that culture requires a different process, Sanfilippo said. "But the good news is it probably can be done."

Promoting Health Services Research

In the Healthcare Innovation Program at Emory University and the Georgia Institute of Technology, it has been a priority to expand health services research and education, Sanfilippo said. Existing programs were scattered across units and institutions, and the resources needed to establish a health services center were limited. The goals of the initiative were to:

- Increase quality, scope, impact, and recognition
- Accelerate interactions across disciplines and units
- Engage other academic and health care organizations
- Develop new activities to accelerate collaboration
- Minimize expense and competition for resources and recognition
- Define measures for assessment and success

For an outlay of about $100,000, the program developed a number of bottom-up projects that were focused on bringing the key individuals together, Sanfilippo said. This included convening collaborations among those in the health system who were involved primarily in quality improvement with investigators in public health, the school of medicine, the school of nursing, the business school, and the school of law. Small seed grants, quarterly symposia, interest groups, research planning, and student–faculty networking nights were among the many steps taken to foster collaboration. Among the results were a significant increase in health services research funding and the engagement of more than 1,700 faculty members and students across more than 50 units, not just from Emory and Georgia Tech, but from other institutions as well.

Sanfilippo drew several lessons from the experience. One should focus on cross-unit and multidisciplinary teams and players, he said, and not duplicate or compete with existing activities. Small investments can yield significant returns, and cost–benefit accounting can justify the use of resources by demonstrating the return on investment and the benefits to patients, students, and staff members. Agreeing on measures of success at the outset is important, as is finding key internal and external advocates.

Alignment is particularly needed among strategic planning (in particular, the quality office), the informatics enterprise, and the research enterprise. "It's a multibody problem," Sanfilippo said.

USING GENOMIC DATA IN THE CLINIC

Many physicians report that they feel unprepared for genomic medicine, said Vassy. Among the many reasons that physicians cite, Vassy highlighted in particular a lack of genomic knowledge, low self-efficacy, EHRs that are not equipped to incorporate genomic information, and a lack of evidence for clinical utility.

The MedSeq Project was designed to explore this unpreparedness by engaging clinicians to use whole-genome sequencing results at Brigham and Women's Hospital. The study involved two groups of 10 primary care providers caring for 100 generally healthy patients in their 40s through 60s and 10 cardiologists overseeing the care of 100 cardiomyopathy patients (see Figure 3-2). "We're studying both the physicians and the patients in this process," Vassy said. Patients in the two groups—the 100 generally healthy patients and the 100 cardiomyopathy patients—

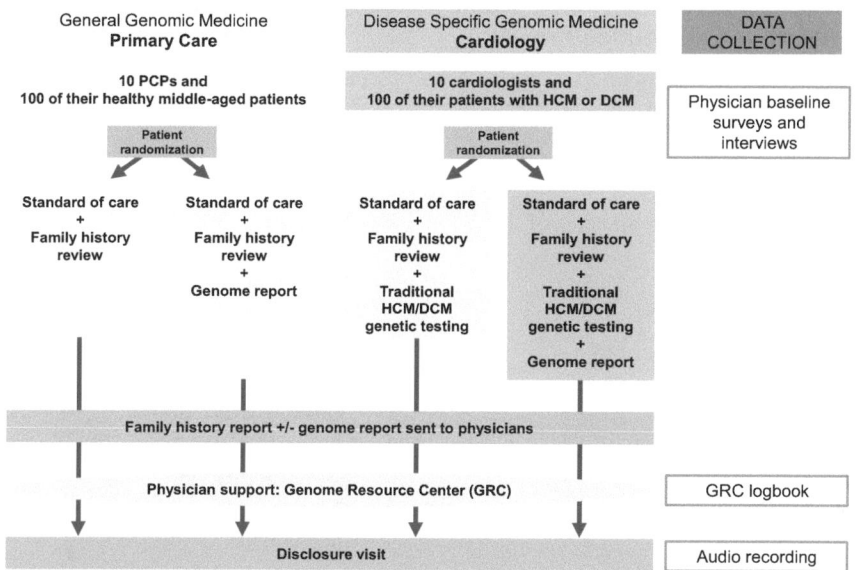

FIGURE 3-2 The MedSeq Project is studying the use of genomic data in the care of 100 healthy patients and 100 patients with hypertrophic cardiomyopathy and dilated cardiomyopathy.
NOTE: DCM, dilated cardiomyopathy; GRC, Genome Resource Center; HCM, hypertrophic cardiomyopathy; PCP, primary care physician.
SOURCE: Vassy, IOM workshop presentation on December 8, 2014.

were randomly assigned to either receive or not receive whole-genome sequencing; family history was reviewed with all patients. The genome reports were sent to the individuals' physicians for review before they discussed the results with their patients. Physicians had access to a Genome Resource Center, where they could speak with medical geneticists, genetic counselors, and other specialists. The conversations between patients and their physicians were audio recorded.

The reports to the physicians of the genome sequencing (lamentably, still delivered in PDF form, Vassy noted) covered monogenic disease risk, carrier status, pharmacogenomics, and blood groups. Physicians also viewed 12 online educational modules that covered a variety of clinical genomic content, each about 15 minutes long (see Table 3-1). "We try to convey this in a brief setting that respects their time and their competing demands," Vassy said.

TABLE 3-1 The 12 Online Genomics Educational Modules for Physicians Offered by the MedSeq Project

Case	Clinical Content Area	Genomic Concepts
1	Familial hypercholesterolemia	• Autosomal dominant and recessive • Modifying genes and penetrance
2	Maturity onset diabetes of the young	• Family history and pedigree analysis • Monogenic forms of common disease
3	Myotonic dystrophy	• Expansion repeat disease and anticipation • Variable expressivity
4	*BRCA*-related disease	• Monogenic forms of common disease • Deletion as a mutation mechanism
5	Alzheimer's disease	• Monogenic forms of common disease • Non-Mendalian genetic risk for common disease
6	Cystic fibrosis	• Autosomal recessive carrier state • Incidental diagnosis of mild disease
7	Hypertrophic cardiomyopathy	• Variants of unknown significance • Database variability
8	Clopidogrel pharmacogenomics	• Cytochrome p450 genetics splice inducing mutations
9	Vascular Ehlers–Danlos syndrome	• Ethical, legal, and social implications of genomic information • Genetic Information Nondiscrimination Act and Massachusetts genetic privacy law
10	Age-related macular degeneration	• Genome-wide association studies and risk
11	Atrial fibrillation	• Management advice in the setting of pre-symptomatic risk
12	Thoracic aortic aneurysm	• Syndromic versus non-syndromic disease

In discussing the results of the study, Vassy focused on the primary care providers. Among the first 10 healthy patients, 3 had a monogenic disease risk, with the three variants being classified as pathogenic, as likely pathogenic, and as of unknown significance with pathogenicity favored. Carrier variants were found in all 10 patients, with a mean of 2.2 such variants per person.

Examples of the questions physicians asked before they talked with patients were:

- Are there standard recommendations for counseling patients concerning the significance of their carrier status for their children?
- Would Ehlers–Danlos syndrome come up on the whole-genome sequence screen? There is a history of this in my patient's family.
- Given that my patient's directed screening for hypertrophic cardiomyopathy genes was negative, are there standard recommendations on the frequency and means for subsequent genetic reassessments?

According to Vassy, physicians did "pretty reasonable things" with the information they received. For example, one physician who was informed of a likely pathogenic variant for Romano–Ward syndrome, a condition that causes irregular heartbeat, ordered an electrocardiogram for the patient.

Physicians also were largely correct in their interpretations of the reports, Vassy said. They took the information from the report and added it to other information they had, such as the results of a physical or family history, to make decisions. They asked questions that the whole-genome sequence report prompted them to ask. They took information from the report and contextualized it for the individual patient. They understood some of the limitations of sequencing, both on the analytic side and regarding clinical decision making.

Physicians who were asked to be part of the study worried about the amount of time it would take, Vassy said, but the researchers tried to integrate the project into the physicians' clinical care processes and not provide it as a separate research visit that was carved out of clinical time. Another issue was how to get genomic information back into the health care system. Providers would document it in their notes, but it might not be structured in a way that would be informative for the whole learning health care system.

In essence, the whole-genome sequencing results were "just like any other clinical test in medicine," Vassy said. The physicians and patients engaged in shared decision making and clinical reasoning, just as they would with other kinds of information. The concept of genetic exceptionalism did not hold in these exchanges.

What was learned from the study, Vassy said, was that in order to help translate genomic sequencing into improved patient outcomes, clinicians need just-in-time information, including test characteristics and limitations, guidelines or expert recommendations for decision making, and time limits on the validity of the information. "Genomic medicine is a rapidly evolving field, and what is true today may not be true 6 months from now, or 2 years, or certainly 5 years from now," he said. The major barrier, he added, is the demonstration that genomics will improve clinical outcomes. Demonstrating the clinical utility of specific genomic tests would also help determine whether the benefits of new information will outweigh the risks for a particular patient and whether the information will change decisions. Clinicians are "going to want to know what difference does this make," he said. "We need to provide that evidence for them before we can expect this to be implemented broadly."

4

Genomics and the EHR in a Learning Health Care System

Important Points Highlighted by the Individual Speakers

- Both academic health centers and community centers are working to incorporate genomic information into their systems, but the efforts are largely separate. Establishing data standards and common ways of representing outcomes would facilitate the scalability of efforts and the translation of genomic information into clinical care. (Moss)
- The most practical way of integrating genomic data into the clinic is to provide it through clinical decision support, but that means the community would need to agree upon common allele and test code nomenclature so that the guidance is scalable and interoperable. (Chute)
- Cultivating a "data donor" culture in which data sharing is commonplace and encouraged because it would help the greater population could be achieved by ensuring the privacy of personal information. (Chute)

Establishing standards for data will facilitate the incorporation of genomic information into the EHR, allow for the interoperability of data flow among systems, and increase the ease with which big data can be shared and managed. Still, even once those data standards have been established, additional challenges to a genomics-enabled EHR will remain, including deciding how and what information will be shared, ensuring equity of access to the information, developing useful clinical decision support and providing clinicians with the knowledge to use it, and providing insurance coverage for genetic tests that have been demonstrated to have clinical value.

LEVERAGING EHRs FOR GENOMICS

The developers of EHRs are working hard to incorporate genomic information into the clinical record and to rapidly translate new discoveries into clinical care, said Moss. The push to incorporate genomic information in the EHR is coming from the consumer demand, he said. It began with the academic medical centers, but now community health centers are also very interested. Much of these efforts are carried out in separate silos, and because of that, efforts are not consistent. Providing consistency would reduce the need for rework in this area, Moss said. Organizations are using genomic information in different ways—some of them to drive alerts, for instance, while other groups are using EHRs in genome-wide association studies.

The disparate approaches, lack of standardization, and limited sharing of approaches all present barriers to developing a scalable genomics-enabled learning health care system. "People can learn best practices from work that others have done," Moss said, "but at a technical level there's no sharing of what's been done to make it easier, especially for the non-academic medical centers that are trying to do this."

Moss pointed to three areas in particular in which changes need to be made in order to make learning health systems a reality. The most important barrier he discussed is the lack of data standards for genomic information. Various standards exist, but they are young and have not been tested and tailored to meet the needs of the genomics community. Getting useful feedback from the genomics community and moving toward standardized data models and exchange formats would help lessen the burden on local efforts to integrate genomic data into care, he said. The DIGITizE Action Collaborative of the Roundtable on Translating Genomic-Based Research for Health (see Chapter 5 for more information), which Epic participates in, is working to solve this issue by assembling a framework for integrating genomic data into the EHR.

A more standardized approach to represent knowledge is also needed, Moss said. For example, standard representations of outcomes data would help make the value proposition for payment models, additional funding, and new research. In addition, the standardization of genomic knowledge in a shareable and scalable way would help speed the translation of discoveries into clinical care. As an example of a system that works well today, Moss cited the system for checking drug–drug interactions, which has been quickly translated into clinical care and is scalable. "This model could work great for something like drug–gene interactions," he said.

Not every system is going to do things in the same way or use the same resources, Moss acknowledged. Already, many different models have been developed to incorporate genomic data into health care, and standards will need to be flexible to support these. Developing those standards will require a collaborative process that crosses many stakeholder groups. The demand for genomic information in EHRs is only growing, he said.

CREATING A SUPPORTIVE INFRASTRUCTURE

A learning health system aligns science and informatics, develops strong patient–clinician partnerships, provides incentives for innovation, and creates a culture of continuous improvement to produce the best care at the lowest cost, said Steve Leffler, the chief medical officer at The University of Vermont Medical Center and a professor of surgery at the University of Vermont College of Medicine. Each of these actions, in the context of genomic medicine, can be used with EHRs to advance health care.

EHRs will need to be optimized to use genomic information effectively, Leffler said. Given that patient charts can become overloaded with extra data, standardizing EHR displays can ensure that health care providers see important information. But, he added, too many alerts can be detrimental because they are eventually ignored.

To personalize care for patients, genomics needs to be incorporated seamlessly into the EHR, Leffler said, and the genomics information needs to be accurate if it is to be useful. Genomic data will be most useful in the background, where they will help providers make good decisions without distracting them from their jobs. Eventually, all health care providers will need to know how to use genomic information, but for now primary care providers who are not comfortable with genomic information can work with genetic counselors, geneticists, pathologists, and others who understand the the test results in order to make genetics-informed health care decisions, he said. Physicians who do not want to learn from computer screens while they are practicing are likely to see this type of expert advice–based learning as a welcome alternative, said Peterson.

Although specialists will be the more likely point of interaction with regard to genomics, Jarvik agreed that "every physician is going to need to become literate in genomic medicine. But we have a long way to go right now." She cited the example of a patient informed of a warfarin sensitivity variant in a research study who was switched to a different drug by a physician. "This drug did not need to be changed as far as we know. We are interviewing the patient and the physician about this experience, and maybe there was some valid reason, but I'm concerned that there wasn't." Health care providers need training to be able to do phenotyping, to see the benefits of genomic information to patients, and to use the information in clinics, said Fowler. "There is a real dearth of skills in this particular area, and for us that's a particular challenge."

There are several barriers to the integration of genomics into the EHR, Leffler said. For example, determining a way to identify who will benefit in the initial stages of integration when not everyone can be included in such a system is a challenge, he said. "Are you going to focus on people who already have a disease, on their family members? Who is going to make those decisions before it's universal?" Informed consent is another issue. Will everyone have to opt in or opt out? Will incidental findings be conveyed to family members who might be affected? Will patients be able to see all their genomic information, including incidental findings? If they are concerned about having a disease, will they be tested for that disease every time? Are providers expected to review every piece of available information? How will the use of genomic data by providers be monitored? Another consideration is that emergencies need to be dealt with quickly, so EHRs cannot slow down responses, Leffler added. "How we're going to deal with incidental findings has to be well understood and planned out ahead of time," he said.

Shared decision making in the age of genomics will also generate challenges. For example, a person in his or her 20s who lacks markers associated with a predisposition to lung cancer may misinterpret the results as meaning that there is protection from the disease and that smoking would be safe. Other patients may not want to know that they are at risk for a disease and would consider such information to be an intrusion into their lives. Another possibility is that patients will be overtested when genomic information is available. The result will be "rich discussions," Leffler said, "but it's going to take a lot of time. You're going to need to have knowledgeable providers who understand that genomics is probabilistic, not deterministic, so these markers can make you more likely to have something, but it's not an absolute." Ultimately, genomics will make possible shared decision making, allow new research, drive improvements in population health, optimize care, and prevent complications, thus driving down the cost of health care and improving value.

The infrastructure for genomic medicine is lacking in critical areas, Jarvik said. For example, not all of the variant annotations are getting into central databases where they can be widely used by academic laboratories and companies. Furthermore, EHRs are not standardized nationally. Information is entered into systems in different ways, the systems do not communicate with each other, and they are not currently standardized to accept genomic information. Institutions often have difficulty communicating with each other because much of their laboratory genetic data are in the form of PDF files, Peterson said, which is "the lowest common denominator to exchange genomic data at this point, and that clearly needs to change."

One thing that would be very helpful, Jarvik said, would be if EHRs automatically pushed these variants to the relevant databases, such as ClinVar. In addition, providing access to all the information from a genetic test, not just the information that goes into a report, could create new opportunities for discovery. Genetic test providers are competing for work, so an incentive could be created for them to adhere to a shared format for complete results, and this incentive could be reinforced with policy.

The transition from the International Classification of Diseases[1] version 9 (ICD-9) to ICD-10 creates a problem for genomic research. Though few codes are fundamentally different, the transition creates discontinuities. For example, a single ICD-9 code can correspond to many ICD-10 codes, and vice versa. "That's intractable in terms of having trivial table lookups," said Chute. Though ICD-11 promises to improve the situation, the current systems are problematic—a point that was reiterated by several other presenters. EHRs need to make it easy for health care providers to do the right thing and hard to make errors, Leffler said. Genomics will add huge amounts of new information to EHRs, and how this information is incorporated and viewed will be critical to how useful it will be, he said.

Clinical Decision Support

Providers need to be knowledgeable about using genomic information and about discussing what the information means with their patients, Leffler said. Providers need better information, not necessarily more data, he continued. If adding genomic information to the EHR only

---

[1] International Classification of Diseases, http://www.who.int/classifications/icd/en (accessed February 23, 2015).

adds data, its usefulness will not be maximized. The key will be to integrate the information in a way that makes sense to providers and adds value to the provider–patient encounter.

The most pragmatic way of taking genomic data and integrating it into the clinical process is through clinical decision support. Computational tools and infrastructure must be available to inform physicians of relevant findings, rather than expecting them to look the finding up or know them off the top of their heads, Chute said. However, he added, the challenge to enabling clinical decision support is that the nomenclature for alleles is collapsing. The number of alleles that must be distinguished is rapidly exceeding the capability of the current system, and the designation of variants is not always consistent. Genetics laboratories are creating their own names and codes for genetic tests, which works against the consistency and comparability of laboratory results. "It's simply not usable for clinical decision support," Chute said.

Decision support tools tend to be binary, yielding yes or no choices, whereas genomics is probabilistic, Leffler said. He also added that for clinical decision support tools to work, the problem list needs to be correct, which is often not the case today.

The classification of variants is also a problem, Jarvik said. The University of Washington has a Return of Results Committee, which has taken on the difficult task of figuring out how to classify challenging variants and deciding what incidental findings should be returned to patients, she said. The committee has identified 112 gene–disease pairs that it considers returnable, along with reporting formats and clinical decision support (Dorschner et al., 2014). When six different genomics laboratories, all of which were certified through the Clinical Laboratory Improvement Amendments, or CLIA, classified the same variants using new guidelines developed by the American College of Medical Genetics, the resulting classifications were the same across labs for only one of the six variants (Amendola et al., 2015). "We're going to have to come up with a system we all agree on," Jarvik said.

Peterson said that health care providers do not always follow the advice provided by clinical decision support, and the reason is often that they have additional information about a patient that factors into their decisions. "We would like our rates of following advice to probably go a little higher than they are," he said, "but it's never going to be 100 percent and probably shouldn't be." The information generated by not following program advice goes back into the EHR, and this information could be used to do comparative effectiveness studies of the value of advice.

MANAGING BIG DATA

Genomic medicine is a big data problem, said Ketan Paranjape, the worldwide director of health and life sciences in the Health Strategy and Solutions Group at Intel Corporation. Large portions of genomic data are accessible, but they are gathered, stored, and disseminated differently in health care than in other industries such as financial services or manufacturing. Furthermore, a variety of types of data exist—not just genomic data, but also clinical trial data, various forms of bioinformatics, and even payer and reimbursement information. Genomic data are being generated not just for individuals but for pathogens, tissues, and other biological entities. Even patients are generating data of various types that could be incorporated into genomic medicine through such means as personal genomic tests and wearable monitors.

Today, the Broad Institute[2] produces amounts of data that are on par with the big cloud producers such as Microsoft, Facebook, and Amazon. Even more data—more than 300 petabytes—is expected to be produced by the Broad Institute in 2015, Paranjape said. And other organizations in the United States and abroad are producing even more data.

Paranjape said that various problems with data generation, management, and interpretation pose barriers to genomic medicine. As a single example, he pointed to the problem of storing genomic data for long periods of time. "Have you thought about keeping the data in your hard disk forever?"

Several projects are intended to overcome these data-related barriers. One is a project of the Charité hospital system in Berlin, which is performing real-time cancer analysis to match patients with the proper therapies. The system uses structured and unstructured data to collect and analyze up to 3.5 million data points per patient, completing in seconds a process that used to take 2 days, Paranjape said. The result has been improved medical care received by patients and provided by doctors and hospitals. Additionally, the system has generated higher-quality information that is usable for research with on-the-fly analysis using medical records, PubMed references, pharmaceutical databases, and survival curve statistics.

The Regional Health Information Network in Jinzhou, China, is another example of an organization that is successfully managing big data.

---

[2]The challenges of analyzing hundreds of thousands of genomes, http://www.broadinstitute.org/~carneiro/talks/20140612-qatar_genomics_conference.pdf (accessed February 25, 2015).

In response to problems involving scalability, performance, maintenance, and data storage, the network developed EHR systems and health care service that run on a distributed computing system in order both to address these issues and to significantly reduce storage costs.

Paranjape cited a system that connects HCC (hierarchical condition category) codes with ICD-9, -10, and -11 codes. The goal is to identify relevant features and patterns behind diseases to more accurately identify suspected conditions in patients.

In addition to developing processors to handle big data, Intel supports several training programs in genomics and technology. Specifically, the company has a team that works with clinicians with the goal of understanding how the clinicians use genomic data. These programs have helped bioinformaticists, life scientists, computer scientists, clinicians, and other professionals work more effectively to create the personalized medicine of the future, Paranjape said.

INSURANCE AND REGULATORY ISSUES

A lack of substantial evidence for the clinical utility of genetic information has led insurers to be reluctant to pay for these tests, Jarvik said. Because of the significant amount of time allocated to consenting patients for testing and then interpreting and explaining the results, covering the costs would make it possible for genetic testing to be implemented in a practical way. The policies of insurers are an obstacle to genomic medicine, Jarvik said. For example, a large insurer in Washington State recently declared that any genetic panel is investigational, including the cystic fibrosis 32-mutation panel. "How does this single-gene test get involved here?" Jarvik asked. "The word 'panel.'" The policy was interpreted in such a way that it did not distinguish between a single-gene and a multiple-gene test, and because the word "panel" was in the test name, it was considered investigational, she said.

Yet insurance coverage is critical as the end point of a process beginning with research and progressing through the development of an evidence base and practice guidelines, Jarvik said. "In medicine, even when we have a lot of evidence of what is best, we still need to get someone to pay for it," she said. "So we have to think about getting practice guidelines from societies based on that evidence in order to convince insurers what is a reasonable level of care to provide for patients." More investments are also needed in outcomes research, she said.

Regulatory changes pose another obstacle for integrating genomics into the health care system. For example, the new patient access rights that CLIA laboratories now have to grant may be interpreted to mean that raw gene variant data files are shared upon request (Evans et al., 2014). This will require that physicians explain the data to their patients, because the laboratories are not required to do so, Jarvik said. New regulations also require that FDA approve tests when variants are deemed to have clinical utility. Some people welcome that oversight, but molecular pathologists are generally not among them, Jarvik said. In general, genetic tests have had few errors, so the public health benefits of this regulation are questionable. "I have patients who have been followed for four generations before we finally solved what was wrong with them because of new technologies, and I don't want to see that limited."

DATA SHARING

Summary level information can be shared nationally and internationally to produce even larger patient cohorts. This type of model is becoming more common, Fowler said, and this means that procedures must be developed to share data and collaborate while protecting the privacy of patients. Making sure that the data are interoperable, so that data from many systems can be aggregated, is an important issue, Chute said.

Peterson made the case for sharing knowledge resources among institutions. In particular, the sharing of knowledge resources is very helpful to programs that are just getting started with performing genomic medicine, including phenotyping algorithms, variant calling, determining the clinical interpretation of variants, and maintaining a rule repository for clinical decision support.

Finally, Chute mentioned the idea of cultivating a "data donor culture." Organ donation is considered popular in the sense that people are proud to tell others about their donor status. But, he said, "there's no coolness being associated with being a data donor, and yet in terms of discovery and integration and learning health systems, nothing is more important culturally than for society to understand the importance of data sharing."

5

Representing Genomic Information in the EHR Ecosystem

Important Points Highlighted by the Individual Speakers

- Transmission of genomic data within the health information ecosystem could be improved by defining data standards that would also foster interoperability and allow for scalability. (Aronson, Nolen)
- The DIGITizE Action Collaborative was established in 2014 to engage key stakeholders for developing a framework for data genomic standards so that this information is more easily integrated into EHR platforms for clinical use. (Aronson, Nolen)
- Setting data standards for genomic information can enable EHR systems—independent of where they are used—and users of these systems to easily understand the data across the medical community. (Nolen)
- Four use cases focused on pharmacogenomics will be the initial starting point for implementing the standards framework designed by the action collaborative. The goal of the effort is to pilot a project and demonstrate how genomic information can effectively flow through a health information technology system. (Aronson)

Connecting health and medical data from genomics to EHRs would be useful to furthering the understanding of disease, increasing the effectiveness of therapies and their safety, and improving health outcomes. However, as several workshop participants pointed out, health information technology systems are not currently represented in a structured, standards-based, and interoperable format. In the effort to use genomic data successfully in the clinic and for research, defining standards for the data would improve interoperability and scalability by providing a common framework that could be understood across systems. Engaging key stakeholders in a collaborative effort to set these standards could improve the ability to integrate data into the EHR. Incorporating those data in a structured format could increase the knowledge gained about health from genomic information.

In 2014, the DIGITizE Action Collaborative[1] was formed as an activity under the auspices of the IOM's Roundtable on Translating Genomic-Based Research for Health. Its aim is to engage key stakeholders from the community in outlining and facilitating a framework for genomic data standards that could be uniformly implemented across health care systems. The action collaborative has brought together representatives of academic health centers, EHR vendors, government, laboratories, standards bodies, and patients to help facilitate the use of genomics in the clinic. The goal is to represent genetic information in a structured format that is interoperable between platforms. The collaborative's members will evaluate their framework for genomic standards by implementing a pilot program to test the flow of information within the EHR ecosystem, which includes health centers, laboratories, and EHRs.

STANDARDS AND SCALE

Today, individual hospitals are using genomics to improve the care of patients, but these individual efforts generally do not scale, said John David Larkin Nolen, a senior director and the general manager of the Laboratory Business Unit for Cerner Corporation. A learning health care system that works on a small scale may not be efficient or effective in a larger facility if the current state of the system is that it cannot handle the rapid growth of genomic knowledge, Nolen said. An effective system needs to work independently of the venue.

Clinicians typically obtain genomic results from a laboratory, perhaps with some guidance about how to interpret those results, Nolen said, but genomics produces too many data for clinicians to rely entirely on their expertise, which creates a need for decision support tools. Using these tools, physicians can navigate through a patient's data, know what to do

---

[1] See http://www.iom.edu/Activities/Research/GenomicBasedResearch/Innovation-Collaboratives/EHR.aspx (accessed March 19, 2015).

with that information, and decide how to take care of that patient—all while new information from research continually flows into the system.

The challenge, Nolen said, is to figure out what data to store, how to transmit and share the data, how to protect patient privacy, and how to be paid so that the benefits of genomics can be realized in any health care setting. One way to address the problem of transmitting large amounts of data among facilities and systems is to create a common vocabulary for the data. Once a standard vocabulary is established, the content of the information can be determined, and this will allow for improved portability of the data and connectivity among data users no matter where they are located. Many different stakeholder groups need to be involved in the establishment of standards, Nolen said, and they need to be guided by a plan on which the various stakeholders have agreed. The action collaborative provides a forum for bringing stakeholders together to decide on a common framework to move genomic data through the EHR ecosystem.

USE CASES

The goal of the action collaborative is to "accelerate the rate at which we develop clinical genomic and information technology support and deploy it," said Sandy Aronson, executive director of information technology for Partners HealthCare Personalized Medicine. By continually evaluating the stakeholder groups represented (government agencies, health care providers, laboratories, information system vendors, standards bodies, patient representatives), the group makes certain that the stakeholders have the right expertise to achieve their goals.

The action collaborative members agreed that their starting point would be use cases that would focus their efforts on moving genomic data through the EHR ecosystem. After taking this small step, the group would still have to address significant challenges to "fundamentally improve patient care," Aronson said, but the defined tasks would be achievable in a short time frame. The group will examine two pharmacogenomics examples (abacavir and Imuran), exploring four specific use cases for each:

- Incorporating genetic results into EHR user interfaces;
- Adding genetic tests in order sets;
- Using clinical decision support to identify when a test should be ordered (pre-test alert); and

- Using clinical decision support to identify when a drug order is inconsistent with a test result (post-test order alert).

The goal of the action collaborative, Aronson said, is not necessarily to publish papers but rather to establish the inter-institutional project management structures needed to deal with the interdependencies in this field (see Figure 5-1). Providers depend on laboratories for data. Laboratories and providers depend on information technology vendors to supply interoperable systems. To meet that requirement, vendors, which are usually competing organizations, depend on each other to establish interfaces among themselves. All stakeholders depend on standards bodies to provide the necessary standards and ontologies to enable interconnectivity to occur. The standards bodies depend on laboratory providers and, to some extent vendors to get the input and models that will make it possible for the clinical decision support to operate robustly.

The funding for generating the genetic data and delivering it through an interface that is interoperable generally flows through providers and laboratories. These entities in turn depend on government agencies to establish a reimbursement framework or to pay for generating and

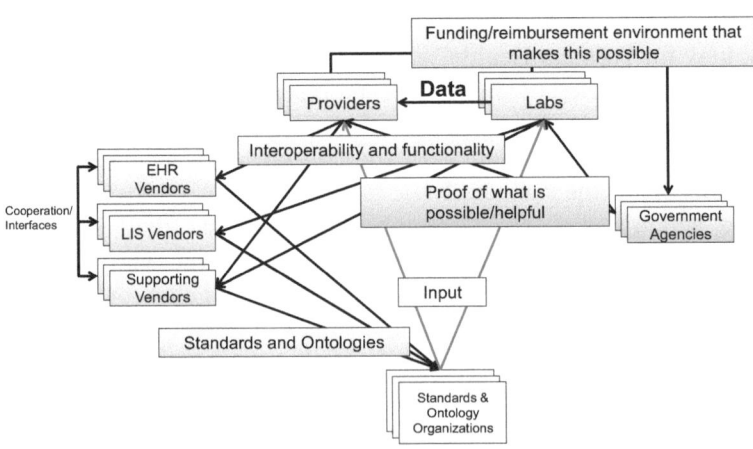

FIGURE 5-1 Interdependencies in the health care system complicate the establishment of inter-institutional project management structures.
NOTE: EHR, electronic health record; LIS, laboratory information system.
SOURCE: Aronson, IOM workshop presentation on December 8, 2014.

delivering the data through grants. Government agencies depend on laboratories and providers to demonstrate that funding will lead to progress. The action collaborative has individuals with expertise in each of these health information system interdependencies, Aronson said.

The group has sought to establish a framework that is detailed enough to enable the different players in this space to make progress efficiently. "Those interdependencies—and the connections between institutions—are what's most holding us back," Aronson said. The scope of genomics is huge, he pointed out, so the action collaborative is trying to focus its efforts on particular parts of the problem so as to make incremental progress. A general theme is to make sure that the data needed for clinical decision support transfer from the laboratory to the provider. "Until that transfer happens, nothing can start, so this is where we're focusing our efforts," he said.

Focusing on a clinical decision support rule would "add value and [allow us to] gain momentum," Aronson said. As an example of how the rule would be used in practice, Aronson described the fourth case study, the post-test alert. In this case, a drug has been ordered for a patient on the basis of a genetic test, and the clinical decision support identifies whether that order is inconsistent with a pharmacogenomics result and alerts the clinician if a problem is detected. One rule that could add value, for example, would be to alert clinicians if they prescribe more than 50 mg of Imuran per day to adult patients who are low metabolizers of the drug. Some false positives will occur with chemotherapy situations, Aronson acknowledged, but enhancements could be added over time.

The action collaborative members decided that additional information needed to be added to the order messages that already are sent between laboratories and providers. For example, one piece of information that needs to be included in the order message is the two words at the top of the test report that summarize the information, such as "abacavir sensitivity," Aronson said, and accompanying those two words would be descriptors such as "high metabolizer" or "low metabolizer." An ontology needs to be established for which words are acceptable to use, he said. The Clinical Pharmacogenetics Implementation Consortium and ClinGen have taken on the task of establishing the ontology, and work with standards bodies will determine how the information will be coded and transported. Aronson said that those involved in the action collaborative, at "both at an organizational level and an individual level, really care about making sure that we deliver the promise of genetic medicine to the patients and the families who could benefit from it."

POTENTIAL NEXT STEPS AND CONSIDERATIONS

The four initial use cases are just a beginning, Nolen said. The action collaborative is starting with pharmacogenomics, but many other issues are waiting. For future projects, the action collaborative will continue to be a coordinating center that keeps projects organized and moving forward. The group will be "the common meeting spot for everyone to come together and help drive this," Nolen said.

In particular, the vendor community can accelerate the process of integrating genomics into the EHR by delivering support to communities. Once the standards exist for vendors' systems to talk with each other, Aronson said, many opportunities will open up to disseminate these systems. "The key is to think big but start small," Nolen said. "While pharmacogenomics might not be that exciting to a lot of people in the room, it's something that's within easy grasp even without firm standards. It's setting up connections, setting up the content, and pushing the data through."

Nolen pointed out that similar problems exist in countries around the world, even where single-payer health systems make coordination easier. Even if the data are connected in a single country, he asked, "how do you [handle] the deluge of data that's coming out of that sequencer? How do you move that into your system in a smart way that scales, that allows you to power up the decision process for your clinicians?" Lessons from other countries' experiences could be valuable in the United States, Aronson said. "There's no reason why knowledge and transaction shouldn't be able to cross international boundaries and discoveries," he said.

It is important to keep the payers engaged in order to get value out of the system, Nolen said. However, in some cases, action needs to occur before the costs of that action will be covered, he said. Reference laboratories, academic hospitals, community hospitals, and other institutions have all been eager to participate in the action collaborative regardless of reimbursement by payers because they realize that there is value in the form of cost savings and in providing better care for their patients. As Aronson said, "The goal of this work should be to make the incorporation of these genetic tests easier and more efficient."

6

Possible Next Steps

Achieving effective integration of genomic data into knowledge-generating health care systems will require interoperability, said Steven Leffler, the chief medical officer at The University of Vermont Medical Center. Adapting current platforms, reusing existing components of systems that work well, and standardizing the structure of the data, including consumer data, will contribute to these efforts. (See Box 6-1 for actions suggested by individual workshop participants for facilitating the integration of genomics into health care systems.) Once the data are in the appropriate format and can be more easily transferred and used, clinical decision support algorithms can be developed to provide the necessary information at the point of care, said Sam Shekar, the chief medical officer within Northrop Grumman's information systems sector. The implementation of genomic data in the clinic is not without its challenges. The introduction of new knowledge into the health care system will likely mean that there will be cultural changes related to how information is used, and it will call for behavioral changes in clinical practice as well, said Geoffrey Ginsburg, director of the Duke University Center for Applied Genomics and Precision Medicine. Ensuring that any changes in health care practice benefit all people and do not introduce unintended disparities in health care will be key, said Alexander Ommaya, the senior director of implementation research and policy at the Association of American Medical Colleges.

BOX 6-1
Possible Next Steps Proposed by Individual Workshop Participants

Interoperability of EHRs
- Ensure that the quality of genomic data is clinical grade and that it is in an accessible format so that it can be used for future research and to inform clinical care. (Risch)
- Support regulations that will make EHRs fully interoperable for genomic information. (Leffler)
- Establish data standards for genomics to allow for EHRs to communicate and for genomic data to flow more easily across labs and systems to providers. (Aronson, Fowler, Nolen)
- To demonstrate how the interoperability of systems can be increased, start with specific health problems whose outcomes are likely to be changed with genomic and other clinical data. (Hill)

Clinical Decision Support
- Reach agreement on allele and test code nomenclature to facilitate the development of clinical decision support tools for genomics. (Chute)
- Create warehouses of clinical decision support tools that can be shared and used widely. (Ginsburg)
- Measure outcomes to determine the validity of algorithms used to guide practice. (Moss)
- Develop a core infrastructure to handle clinical decision support and the long-term storage of complex data. (Nolen)

Data Sharing
- Build platforms with reusable components that are scalable and can be implemented anywhere. (Friedman)
- Standardize data so that they can be re-used. (Chute)
- Foster interoperable health care systems to enable genomic data sharing. (Terry)
- Inform the public about data sharing to cultivate a "data donor" culture. (Chute)
- Network data from around the world to increase the size of databases and power of research studies. (Aronson)
- Integrate patient-provided data into health care information technology systems. (Baker)
- Examine whether personally controlled health databanks can make data accessible for sharing while protecting privacy. (Friedman)
- Support research to understand and generate personalized user interfaces and preferences. (Baker)

Implementation
- Engage groups with a particular interest and who value genomics, such as people with undiagnosed or chronic diseases, to demonstrate the full potential of this information. (Terry)
- Measure and track health and health care disparities to determine the impact of genomics-based interventions. (Ommaya)
- Support social science and behavioral research to understand the priorities and values of patients and providers when genomics is introduced in the clinic. (Ginsburg)

EHR INTEROPERABILITY

To share information seamlessly, EHRs need to be fully interoperable for genomic information and other clinical information, Lcfflcr said. Purchasers of EHR systems can demand that vendors provide this feature, though they have much less leverage if they already have bought a proprietary system. Regulatory bodies also can help push for interoperability, as can entities like the DIGITizE Action Collaborative.

It is important to establish a common set of standards across EHR platform versions and the providers who use them, said Tom Fowler, the director of public health at Genomics England. Currently, not even systems from the same vendor can easily communicate. Standards for data representations, problem lists, medication lists, and other features will make the EHR a more useful tool. The Department of Veterans Affairs has faced challenges in coordinating all of the various forms of data, a participant said. Genome-wide association studies are being conducted on approximately one-third of 1 million veterans; about 25,000 of the samples have undergone exome sequencing, and roughly 2,000 samples have undergone whole-genome sequencing. There is a large amount of data being generated, but the difficulty has been standardizing the information so that it can be studied optimally. Such standards would greatly help coordinating data in different forms.

Adapting platforms built in recent years is one way to take on current problems, said Charles Friedman, the Josiah Macy Jr. Professor and chair of the Department of Learning Health Sciences at the University of Michigan Medical School. For example, establishing health databanks could be a scalable approach to sharing data while protecting privacy. Scalability of platforms is needed, Friedman said, and this can be facilitated by reusing components of systems elsewhere. Just as a bankcard can work in many different ATMs because people demanded interoperability, widespread public demand for interoperable health care systems could produce change, said Sharon Terry, the president and chief executive officer of the Genetic Alliance. Health databases, for example, could be scaled and maintained for both data sharing capabilities and maintaining privacy. Other potentially scalable programs include the SMART® platform[1] (substitutable apps that can be integrated with EHR systems), a data normalization pipeline for phenotyping called PhenoTips,[2] and the Clinical Decision Support Consortium.[3] "These are just a few examples of platforms that we might be able to incorporate into things that we are doing, saving years, literally, of work," Friedman said. Ginsburg added that perhaps these software platforms and clinical decision support tools could be put into a shareable warehouse for dissemination. The public also could become more involved in specific policy issues, such as policies on genomic data sharing, Terry said.

Instead of trying to fix the problems with the current EHR platforms, what may be needed is to "create a new [EHR] system that would be universal across all the different health care systems," suggested Debra Leonard, a professor and the chair of pathology and laboratory medicine at The University of Vermont Medical Center. There would be significant resistance to discarding the current system because of the large amount of investments that organizations have made in it, said Friedman and Fred Sanfilippo, the director of the Emory–Georgia Tech Healthcare Innovation Program. An alternative would be to incorporate platforms that add value into the existing legacy systems, which is why standards are needed to ensure the interoperability of these augmented systems, said Sandy Aronson, the executive director of information technology for Partners HealthCare Personalized Medicine.

By addressing interoperability issues with two or three specific health problems in which there is an argument that genomic and other clinical data can produce game-changing outcomes in a short time, it should be possible for the push for interoperability to gain traction, said Colin Hill, the chief executive officer of GNS Healthcare. This will be particularly true where there are economic incentives. "It's no accident that oncology is one of the first places where you're seeing a lot of data sharing," Hill said. Oncology and the prevention of preterm births are

---

[1]SMART, http://smartplatforms.org (accessed March 2, 2015).
[2]PhenoTips, https://phenotips.org (accessed March 2, 2015).
[3]Clinical Decision Support Consortium, http://www.cdsconsortium.org (accessed March 2, 2015).

two examples of areas where patients, providers, and payers are all interested in making progress with genomics. "Some of those areas are emerging, but we need to be thoughtful and careful about the health economics of those areas up front," he said. "At the end of the day the economics have to make sense." For many diseases, said Lynn Etheredge of the Rapid Learning Project, the important questions will be how much genetic information is needed, for which patients, and who is going to pay for generating that information.

CLINICAL DECISION SUPPORT

Genomics is at the forefront of clinical decision support because the data are inherently computable and therefore could be provided as supporting information at the point of care, a workshop participant said. Clinical decision support provides a better opportunity for physicians and patients to use genomic data than either could have when the information is solely contained in paper format. Validating the algorithms that are being used to guide practice will be important, Ginsburg said. Aronson said that today's in silico prediction algorithms are "extremely noisy," so that "you wouldn't want to base a significant clinical decision on one of those algorithms alone." Validating these algorithms will require measuring outcomes, said Scott Moss, who leads the research informatics research-and-development team at Epic.

Market forces could lead to algorithms from different vendors being compared to see what effect an intervention had, said Andrew Kasarskis, the co-director of the Icahn Institute for Genomics and Multiscale Biology at Mount Sinai Hospital. "To the extent that an organization has a good handle on what its costs are and somewhat structured data on its clinical population, you'll be able to do some comparisons," he said. In this case, good data standards would further market comparisons and improve health.

Machine learning is one possible way to produce results to guide clinical decision support. But Ketan Paranjape, the worldwide director of health and life sciences in the Health Strategy and Solutions Group at Intel Corporation, said he was worried about decisions that are based on inadequate databases. Even 50,000 medical records in a system are not enough, he warned. One way to increase the size of databases is to network information from around the world, Aronson noted.

FDA may move to regulate decision support systems, said John David Larkin Nolen, the managing director of laboratory medicine at Cerner Corporation. This is another argument for interoperability, he said. "We have to start talking about core infrastructures because that will make it a more sustainable project, versus a one-off here, a one-off there." A core infrastructure could handle both the long-term storage of complex data, which does not necessarily belong in an EHR, and decision support.

UNDERSTANDING CONSUMER VALUE AND PREFERENCES

The cultural issues associated with genomics-enabled health care systems have been studied less than the technology issues, Terry said. The consumers of health care will not engage in something unless it offers value to them. One way to start engaging the public, said Terry, would be to start with a subset of people where the value is high, such as those with undiagnosed or chronic diseases and then the reach could be broadened to a larger population. "We're very engaged in [activities] like comparing plumbers or figuring out the best car to buy, because we have tools to do it and it's important to us," she said. The banking and auto industries have focused on designing their products based on what people want because there is a demand from consumers. We need to "activate the public" to become more informed about genomics and create a demand around individual needs, but in order to accomplish that, the right tools need to be available.

Dixie Baker, a senior partner at Martin, Blanck & Associates, noted earlier that individuals provide the information that drives a knowledge-generating health care system and that understanding their preferences for data sharing is key. The value of personalized user interfaces will require research to understand and generate, she said. "We're trying to engage everyone, and to do that we have to recognize the diversity of people's values." As an example of this kind of work, Paranjape noted that he and his colleagues worked with medical ethnographers to figure out what representation of data on a computer screen would be most effective.

An increasingly important issue will be how to accommodate and integrate patient-provided data, whether from wearable monitors, home sensors, or personal genomic tests, into health care information technology systems, said Baker. "Consumers are definitely investing their own dollars in health products and services," she said. "This is a very rich area of development and research."

Moss observed that consumer health data are already flowing into EHRs at many U.S. organizations, and the trend is accelerating, especially as personal monitors become more common. The consumers decide what data go where and who will have access to the data, with control "very much in the hands of the consumer," he said.

Although there are some concerns about the quality of such consumer-provided data, their potential is great, Paranjape said. As an example, he mentioned a wearable device that can track the tremors of Parkinson's patients and detect changes caused by medications. And computing will continue to become ever more powerful, which will enable data analysis to be more distributed than it is today, Fowler said. People will be analyzing their own genomic information and coming to the health care system to discuss what they have found and, in some cases, to seek treatment.

The extent to which health data will be open remains an unanswered question, Friedman said. As an example, he mentioned personally controlled health databanks, in which an individual contracts with a health databank to be the custodian of his or her health data (see Chapter 3 for a discussion on PEER). "What goes in is what that person wants to go in … and what gets released is just what that person wants released," he said. The idea is completely scalable, he added, and the data are open if the patient agrees that the data can be used for a specific purpose. "There can be multiple banks, just as there are multiple financial banks, competing with each other, but all providing the same services using the same standards. I think that's a very important concept that this community should be aware of." Aronson commented that such a system also would be useful when a patient moved from one health care system to another.

People may not want all of their health information in a personal health databank. For example, they may not want the databank to contain a record of sensitive health issues from when they were young. But Friedman pointed out that the data in a health bank would not be the only instantiation of the data. A health system could also retain data in an EHR, so people could choose which data to retain in a data bank. Baker added that data coming to a bank would need to be digitally signed to ensure the integrity of the data.

Accounting for individual preferences may require applications of social science and behavioral research. Ommaya said that the Association of American Medical Colleges is involving the implementers of health care systems in the entire development process, from the formulation of questions to the conduct of studies to the dissemination and implementation of results. People need to be acculturated to new systems if they are to be comfortable with those systems and use them, Ommaya said.

Friedman agreed that translating new knowledge into practice requires psychology, communication science, implementation science, and other behavioral sciences. Sanfilippo remarked on the number of M.D./Ph.D. programs that are considering Ph.D.s in the social sciences because of the importance of behavioral research and related disciplines. "That's a trend that you're going to see accelerate," he predicted. Behavioral change involves incentives, disincentives, rewards, and recognition, Sanfilippo added, all of which are being studied by the field of behavioral economics.

CLOSING REMARKS

Aligning efforts to integrate genomics into the health care system and to use the information effectively will require an understanding of the priorities and values of patients and providers, Ginsburg said. We would like to see action come from this meeting. At the end of the day, he said, we should be thinking about how we can "begin to build this system that is going to support genomics-enabled health care."

References

Amendola, L. M., M. O. Dorschner, P. D. Robertson, J. S. Salama, R. Hart, B. H. Shirts, M. L. Murray, M. J. Tokita, C. J. Gallego, D. S. Kim, J. T. Bennett, D. R. Crosslin, J. Ranchalis, K. L. Jones, E. A. Rosenthal, E. R. Jarvik, A. Itsara, E. H. Turner, D. S. Herman, J. Schleit, A. Burt, S. M. Jamal, J. L. Abrudan, A. D. Johnson, L. K. Conlin, M. C. Dulik, A. Santani, D. R. Metterville, M. Kelly, A. K. Foreman, K. Lee, K. D. Taylor, X. Guo, K. Crooks, L. A. Kiedrowski, L. J. Raffel, O. Gordon, K. Machini, R. J. Desnick, L. G. Biesecker, S. A. Lubitz, S. Mulchandani, G. M. Cooper, S. Joffe, C. S. Richards, Y. Yang, J. I. Rotter, S. S. Rich, C. J. O'Donnell, J. S. Berg, N. B. Spinner, J. P. Evans, S. M. Fullerton, K. A. Leppig, R. L. Bennett, T. Bird, V. P. Sybert, W. M. Grady, H. K. Tabor, J. H. Kim, M. J. Bamshad, B. Wilfond, A. G. Motulsky, C. R. Scott, C. C. Pritchard, T. D. Walsh, W. Burke, W. H. Raskind, P. Byers, F. M. Hisama, H. Rehm, D. A. Nickerson, and G. P. Jarvik. 2015. Actionable exomic incidental findings in 6503 participants: Challenges of variant classification. *Genome Research* 25(3):305–315.

Bielinski, S. J., J. E. Olson, J. Pathak, R. M. Weinshilboum, L. Wang, K. J. Lyke, E. Ryu, P. V. Targonski, M. D. Van Norstrand, M. A. Hathcock, P. Y. Takahashi, J. B. McCormick, K. J. Johnson, K. J. Maschke, C. R. Rohrer Vitek, M. S. Ellingson, E. D. Wieben, G. Farrugia, J. A. Morrisette, K. J. Kruckeberg, J. K. Bruflat, L. M. Peterson, J. H. Blommel, J. M. Skierka, M. J. Ferber, J. L. Black, L. M. Baudhuin, E. W. Klee, J. L. Ross, T. L. Veldhuizen, C. G. Schultz, P. J. Caraballo, R. R. Freimuth, C. G. Chute, and I. J. Kullo. 2014. Preemptive genotyping for personalized medicine: Design of the right drug, right dose, right time—Using genomic data to individualize treatment protocol. *Mayo Clinic Proceedings* 89(1):25–33.

Delaney, J. T., A. H. Ramirez, E. Bowton, J. M. Pulley, M. A. Basford, J. S. Schildcrout, Y. Shi, R. Zink, M. Oetjens, H. Xu, J. H. Cleator, E. Jahangir, M. D. Ritchie, D. R. Masys, D. M. Roden, D. C. Crawford, and J. C. Denny. 2012. Predicting clopidogrel response using DNA samples linked to an electronic health record. *Clinical Pharmacology and Therapeutics* 91(2):257–263.

Denny, J. C., L. Bastarache, M. D. Ritchie, R. J. Carroll, R. Zink, J. D. Mosley, J. R. Field, J. M. Pulley, A. H. Ramirez, E. Bowton, M. A. Basford, D. S. Carrell, P. L. Peissig, A. N. Kho, J. A. Pacheco, L. V. Rasmussen, D. R. Crosslin, P. K. Crane, J. Pathak, S. J. Bielinski, S. A. Pendergrass, H. Xu, L. A. Hindorff, R. Li, T. A. Manolio, C. G. Chute, R. L. Chisholm, E. B. Larson, G. P. Jarvik, M. H. Brilliant, C. A. McCarty, I. J. Kullo, J. L. Haines, D. C. Crawford, D. R. Masys, and D. M. Roden. 2013. Systematic comparison of phenome-wide association study of electronic medical record data and genome-wide association study data. *Nature Biotechnology* 31(12):1102–1110.

Dorschner, M. O., L. M. Amendola, B. H. Shirts, L. Kiedrowski, J. Salama, A. S. Gordon, S. M. Fullerton, P. Tarczy-Hornoch, P. H. Byers, and G. P. Jarvik. 2014. Refining the structure and content of clinical genomic reports. *American Journal of Medical Genetics Part C: Seminars in Medical Genetics* 166C(1):85–92.

Etheredge, L. M. 2014. Rapid learning: A breakthrough agenda. *Health Affairs (Millwood)* 33(7):1155–1162.

Evans, B. J., M. O. Dorschner, W. Burke, and G. P. Jarvik. 2014. Regulatory changes raise troubling questions for genomic testing. *Genetics in Medicine* 16(11):799–803.

Friedman, C., J. Rubin, J. Brown, M. Buntin, M. Corn, L. Etheredge, C. Gunter, M. Musen, R. Platt, W. Stead, K. Sullivan, and D. Van Houweling. 2015. Toward a science of learning systems: A research agenda for the high-functioning learning health system. *Journal of the American Medical Informatics Association* 22(1):43–50.

Ginsburg, G. 2014. Medical genomics: Gather and use genetic data in health care. *Nature* 508(7497):451–453.

IOM (Institute of Medicine). 2012. *Best care at lower cost: The path to continuously learning health care in America.* Washington, DC: The National Academies Press.

Keroack, M. A., N. R. McConkie, E. K. Johnson, G. J. Epting, I. M. Thompson, and F. Sanfilippo. 2011. Functional alignment, not structural integration, of medical schools and teaching hospitals is associated with high performance in academic health centers. *American Journal of Surgery* 202(2):119–126.

Lauer, M. S., and D. Bonds. 2014. Eliminating the "expensive" adjective for clinical trials. *American Heart Journal* 167(4):419–420.

Lauer, M. S., and R. B. D'Agostino, Sr. 2013. The randomized registry trial—The next disruptive technology in clinical research? *New England Journal of Medicine* 369(17):1579–1581.

Ledford, H. 2013. "Master protocol" aims to revamp cancer trials. *Nature* 498(7453):146–147.

Ray, T. 2013. Two conflicting prospective RCTs on warfarin PGx provide no definitive guidance to physicians. *GenomeWeb.* https://www.genomeweb.com/clinical-genomics/two-conflicting-prospective-rcts-warfarin-pgx-provide-no-definitive-guidance-phy (accessed March 20, 2015).

Ray, T. 2015. VA launches precision oncology program in New England with hopes of future national expansion. *GenomeWeb.* https://www.genomeweb.com/sequencing-technology/va-launches-precision-oncology-program-new-england-hopes-future-national (accessed March 27, 2015).

Ritchie, M. D., J. C. Denny, D. C. Crawford, A. H. Ramirez, J. B. Weiner, J. M. Pulley, M. A. Basford, K. Brown-Gentry, J. R. Balser, D. R. Masys, J. L. Haines, and D. M. Roden. 2010. Robust replication of genotype–phenotype associations across multiple diseases in an electronic medical record. *American Journal of Human Genetics* 86(4):560–572.

Robinson, P. N. 2012. Deep phenotyping for precision medicine. *Human Mutation* 33(5):777–780.

Sanfilippo, F., N. Bendapudi, A. Rucci, and L. Schlesinger. 2008. Strong leadership and teamwork drive culture and performance change: Ohio State University Medical Center, 2000–2006. *Academic Medicine* 83(9):845–854.

Starren, J., M. S. Williams, and E. P. Bottinger. 2013. Crossing the omic chasm: A time for omic ancillary systems. *JAMA* 309(12):1237–1238.

Steinberg, G. B., B. W. Church, C. J. McCall, A. B. Scott, and B. P. Kalis. 2014. Novel predictive models for metabolic syndrome risk: A "big data" analytic approach. *American Journal of Managed Care* 20(6):e221–e228.

Terry, S. F., R. Shelton, G. Biggers, D. Baker, and K. Edwards. 2013. The haystack is made of needles. *Genetic Testing and Molecular Biomarkers* 17(3):175–177.

Toh, S., M. E. Reichman, M. Houstoun, M. Ross Southworth, X. Ding, A. F. Hernandez, M. Levenson, L. Li, C. McCloskey, A. Shoaibi, E. Wu, G. Zornberg, and S. Hennessy. 2012. Comparative risk for angioedema associated with the use of drugs that target the renin-angiotensin-aldosterone system. *Archives of Internal Medicine* 172(20):1582–1589.

Wall, J. D., L. F. Tang, B. Zerbe, M. N. Kvale, P. Y. Kwok, C. Schaefer, and N. Risch. 2014. Estimating genotype error rates from high-coverage next-generation sequence data. *Genome Research* 24(11):1734–1739.

A
Workshop Agenda

Genomics-Enabled Learning Health Care Systems: Gathering and Using Genomic Information to Improve Patient Care and Research: A Workshop
December 8, 2014

The Keck Center of the National Academies, Room 100
500 Fifth Street, NW
Washington, DC 20001

Meeting Objectives

- To explore how key pieces of genetic/genomic information can be effectively and efficiently delivered to patients and clinicians for improving care.
- To discuss how both the health care system and genomic data can be used for evidence generation in research and in patient care.
- To assess current best practices for using knowledge-generating/learning health care systems and which models may provide an opportunity for genomics to be used in the rapid learning process.

Agenda

8:30–8:35 a.m.   **Welcoming Remarks**

Sharon F. Terry, *Roundtable Co-Chair*
President and Chief Executive Officer
Genetic Alliance

Geoffrey Ginsburg, *Roundtable Co-Chair*
Director, Center for Applied Genomics & Precision Medicine; Professor of Medicine and of Pathology and Biomedical Engineering
Duke University

8:35–8:45 a.m.   **Charge to Workshop Speakers and Participants**

Sam Shekar, *Workshop Co-Chair*
Chief Medical Officer
Northrop Grumman Health IT

Geoffrey Ginsburg, *Workshop Co-Chair*
Director, Center for Applied Genomics & Precision Medicine; Professor of Medicine and of Pathology and Biomedical Engineering
Duke University

Session I: Introduction—Opportunities for a Learning Health Care System

Objectives: To understand the advantages of implementing a learning system, the potential economic incentives, and how genomic information may fit into this model.

8:45–9:05 a.m.   Lynn Etheredge
Director
Rapid Learning Project

Session II: Advancing Patient Care and Research with Genomic Information

Objectives: To explore multiple sources of genomic data within the health care system and how it can be captured, accessed, and used as evidence for advancing patient care and research.

Moderator: Debra Leonard, University of Vermont College of Medicine

9:05–9:35 a.m.	**Introduction: Ensuring Quality of Genomic Data**
	Neil Risch Professor, Division of Biostatistics; Director, Institute for Human Genetics University of California, San Francisco; Adjunct Investigator Kaiser Permanente Division of Research

	Josh Peterson
	Assistant Professor of Biomedical
	   Informatics, Assistant Professor of
	   Medicine; Director of Health
	   Information, Technology Evaluation
	Vanderbilt University Medical Center

9:35–10:05 a.m.	**Advancing Research**
	Chris Chute Section Head of Medical Informatics, Professor of Medical Informatics Mayo Clinic
	Tom Fowler Director of Public Health Genomics England
10:05–10:20 a.m.	**Break**

10:20–10:50 a.m.	**Advancing Patient Care**
	Gail Jarvik Head and Professor, Division of Medical Genetics University of Washington School of Medicine
	Stephen Leffler Chief Medical Officer University of Vermont Medical Center
10:50–11:45 a.m.	**Discussion with Speakers and Attendees**
	Chris Chute Tom Fowler Gail Jarvik Stephen Leffler Josh Peterson Neil Risch
11:45 a.m.–12:35 p.m.	**Working Lunch**

Session III: Translation of Genomics for Patient Care and Research

Objectives: To examine how to build a knowledge-generating health care system for genomics and incentivize its use.

Moderator: Andrew Kasarskis, Icahn School of Medicine at Mount Sinai

12:35–1:35 p.m.	**Knowledgeable Health Care System for Genomics**
	Introduction/Overview
	Chuck Friedman Josiah Macy Jr. Professor of Medical Education

Chair, Department of Learning Health Sciences, Medical School
Professor of Information and Public Health
University of Michigan

Health System Data Research Perspective

Colin Hill
Chief Executive Officer, President, Chairman, and Co-Founder
GNS Healthcare

Health System Perspective

Alexander Ommaya
Senior Director, Clinical Effectiveness and Implementation Research
Association of American Medical Colleges

Fred Sanfilippo
Director, Healthcare Innovation Program
Emory University–Georgia Tech

1:35–2:35 p.m. **Clinical End User**

Jason Vassy
Section of General Internal Medicine, VA Boston Healthcare System
Division of General Medicine
Brigham and Women's Hospital;
Instructor
Harvard Medical School

Patient Perspective

Dixie Baker
Senior Partner
Martin, Blanck & Associates

Commercial Developer Perspective

Scott Moss
Research Informatics Software
  Developer
Epic

Beyond the EHR

Ketan Paranjape
Director, Personalized Medicine
Intel

2:35–2:50 p.m. **Break**

2:50–3:50 p.m. **Discussion with Speakers and Attendees**

Dixie Baker
Chuck Friedman
Scott Moss
Alexander Ommaya
Ketan Paranjape
Fred Sanfilippo
Jason Vassy

Additional Discussant:
Sandy Aronson
Director of Information Technology
Partners Healthcare Personalized
  Medicine

Session IV: Integrating Genomic Information into the EHR

Objectives: To examine ongoing efforts to develop a framework and guidance for how genomic information may effectively be integrated and used within the health care system workflow.

Moderator: Sam Shekar, Northrop Grumman Health IT

3:50–4:10 p.m.	**Call to Action** Sandy Aronson Director of Information Technology Partners Healthcare Personalized Medicine John David Nolen Managing Director, Laboratory Business Unit Cerner Corporation
4:10–4:35 p.m.	**Discussion with Speakers and Attendees**

Session V: Considerations for the Future

What are the next steps for achieving effective integration of genomic data into the health care system for the short term and long term? What are the challenges associated with this incorporation, and how can they be overcome to ensure that the information is fully utilized for improving patient care and research? Are there economic or other incentives for health care systems to invest in the knowledge-generating model?

Moderator: Geoffrey Ginsburg, Duke University

4:35–5:30 p.m.	**Discussants** Lynn Etheredge Tom Fowler Chuck Friedman Colin Hill Stephen Leffler John David Nolen Sharon F. Terry
5:30–5:35 p.m.	**Summary and Concluding Remarks** Sam Shekar, *Workshop Co-Chair* Chief Medical Officer Northrop Grumman Health IT

Geoffrey Ginsburg, *Workshop Co-Chair*
Director, Center for Applied Genomics
 & Precision Medicine; Professor of
 Medicine and of Pathology and
 Biomedical Engineering
Duke University

5:35 p.m. **Adjourn**

B

Speaker Biographical Sketches

Sandy Aronson, **M.A.**, is the executive director of information technology for Partners HealthCare Personalized Medicine. His team develops the information technology infrastructure required to support the evolution and practice of genetics-based personalized medicine in both patient-facing and laboratory settings. The system ecosystem that the team maintains enables a real-time continuous learning process that allows laboratories to harness their clinical testing flows to advance the knowledge surrounding genetic variation. The infrastructure makes it possible for laboratories to keep treating clinicians up to date through patient specific alerts as the laboratories improve variant classifications. Mr. Aronson's team developed and continuously enhances the GeneInsight Suite (GIS). GIS consists of GeneInsight Lab, GeneInsight Clinic, and the functionality that underlies the GeneInsight Network, including the VariantWire share-and-share-alike network. The team works with Partners HealthCare Information Systems to maintain the integration between GeneInsight Clinic and the Partners HealthCare electronic health record. In addition to GeneInsight, the team developed the original GIGPAD system to support laboratory processing, including its case management system component; it is developing an Olego management system; and it assists the Partners HealthCare Personalized Medicine Bioinformatics Team in the development and maintenance of bioinformatic pipeline infrastructure.

Previously, Mr. Aronson was an information technology consultant to the biotechnology industry, working for Tribiosys. Mr. Aronson also held several positions with Sapient Corporation, was a strategic consultant for Monitor Company, and founded LearningAction, a Web-based training company now part of Best Software. Mr. Aronson holds a master's degree in organizational behavior and a bachelor's degree in computer science from Stanford University. He also holds a master's degree in biology from Harvard Extension School.

Dixie Baker, Ph.D., is a senior partner at Martin, Blanck & Associates, where she provides consulting services in the areas of health information technology (HIT), electronic health records, privacy and security, and the sharing and protection of genomic data. As a consultant to the Genetic Alliance and its Patient Powered Research Network project, she is participating as a member of the Data Standards, Security, and Network Infrastructure Task Force of the Patient-Centered Outcomes Research Institute National Clinical Research Network project. Since 2009, Dr. Baker has served as a member of the HIT Standards Committee (HITSC), a federal advisory committee created by the American Recovery and Reinvestment Act to recommend standards, implementation specifications, and certification criteria for electronic health record technology. Currently serving her second term, Dr. Baker chairs the HITSC Transport and Security Workgroup, and she previously chaired the Nationwide Health Information Network power team, which developed metrics for determining when a technology standard is ready to be considered as a national standard. Dr. Baker co-chairs the Security Working Group of the Global Alliance for Genomics and Health, a coalition to enable the sharing of genomic data among more than 40 countries. She is a fellow of the Healthcare Information and Management Systems Society. In 2010, Dr. Baker was selected as one of the "Federal 100" top executives from government, industry, and academia who had the greatest impact on the government information systems community. In 2013, www.health careinfosecurity.com selected her as one of its inaugural "Top 10 Influencers in Health Information Security."

Christopher Chute, M.D., Dr.P.H., is section head and a professor of medical informatics at the Mayo Clinic in Rochester, Minnesota. He received his undergraduate and medical training at Brown University, his internal medicine residency at Dartmouth, and his doctoral training in epidemiology at Harvard University. He is board certified in internal medicine and clinical informatics, and is a fellow of the American College of Physicians, the American College of Epidemiology, and the American College of Medical Informatics. He became founding chair of the Division of Biomedical Informatics at Mayo in 1988 and served as division chair for 20 years. He is a principal investigator on Mayo's

Clinical and Translational Science Awards (CTSA) informatics core, the eMERGE cooperative agreement on genotype-to-phenotype association, the Pharmacogenomics Research Network Ontology Resource, the LexGrid projects, and a co–principal investigator on the National Center for Biomedical Ontology. Recent grants as principal investigator include the Department of Health and Human Services/Office of the National Coordinator (ONC) SHARP (Strategic Health IT Advanced Research Projects) on Secondary EHR Data Use and the ONC Beacon Community (co–principal investigator). Dr. Chute is the medical director of health information management at Mayo, chairs data governance, and serves on Mayo's enterprise Information Technology Oversight Committee, and CTSA executive committee. He is presently the chair of the ISO Health Informatics Technical Committee (ISO TC215), and he also chairs the World Health Organization ICD-11 Revision. He serves on the HL7 advisory board. Recently held positions include service as an index member on the Health Information Technology Standards Committee for ONC in the U.S. Department of Health and Human Services and the initial chair of the Biomedical Computing and Health Informatics study section at the National Institutes of Health.

Lynn Etheredge currently heads the Rapid Learning Project. His career started at the White House Office of Management and Budget (OMB), where he was OMB's principal analyst for Medicare and Medicaid and led its work on national health insurance proposals. Mr. Etheredge headed OMB's professional health staff in the Carter and Reagan administrations. He is a founding member of the National Academy of Social Insurance. His contributions have ranged broadly across Medicare, Medicaid, health insurance coverage, managed competition in health care, retirement and pension policies, budget policy, and information technology.

Mr. Etheredge proposed the concept of the "rapid-learning health system" in a special issue of *Health Affairs* in 2007 and is collaborating widely in developing this approach. Rapid-learning initiatives are now generating comparative effectiveness research, a national system of learning networks and research registries covering more than 150 million patients, national biobanks with linked electronic health record and genomic data, a new Medicare and Medicaid Innovation Center (with $10 billion of funds), rapid-learning systems for cancer care and pediatrics, and a new National Science Foundation–supported science of learning systems.

He serves on the editorial board of *Health Affairs* and is author of more than 85 publications. He is a graduate of Swarthmore College.

Tom Fowler, Ph.D., is currently on assignment from Public Health England (PHE) to Genomics England as its director of public health. Dr. Fowler is working to support the science stream around rare diseases, infectious diseases, and cancer. He is also a consultant in Epidemiologist PHE's Field Epidemiology Services, where his focus is infectious disease control. He is an honorary research fellow in public health in the University of Birmingham. He was editor-in-chief of the 2011 Volume 1 of the Chief Medical Officer of England's Annual Report and co-editor of Volume 2. The first of these volumes was a comprehensive review of health data for England, and the second volume was an assessment of the future challenges raised by infectious diseases and antibiotic resistance. Dr. Fowler continues to work with the chief medical officer of England on thought leadership around public health, including well-being and antimicrobial resistance. He is also active in the public health genomics community and was a member of the Public Health Genomics European Network meeting that led to the Declaration of Rome 2012 summary of the European Best Practice Guidelines for Quality Assurance, Provision and Use of Genome-Based Information and Technologies.

Charles Friedman, Ph.D., is the Josiah Macy Jr. Professor and chair of the Department of Learning Health Sciences at the University of Michigan Medical School. He joined the University of Michigan in September 2011 as a professor of information and public health and the director of the University of Michigan health informatics program. Throughout his career, Dr. Friedman's primary academic interests have combined biomedical and health informatics with the processes of education and learning. Dr. Friedman's department is a "first in the nation" medical school academic department dedicated to the sciences of learning at all levels from scale: from learning by individuals to learning by teams and organizations and to learning by ultra-large-scale systems such as entire nations.

Before coming to the University of Michigan, Dr. Friedman held executive positions at the Office of the National Coordinator for Health Information Technology (ONC) in the U.S. Department of Health and Human Services: from 2007 to 2009 as deputy national coordinator, and from 2009 to 2011 as ONC's chief scientific officer. While at ONC, Dr. Friedman oversaw a diverse portfolio which included early steps toward the development of a nationwide learning health system, the education of the nation's health information technology (HIT) workforce, research to improve HIT, program evaluation, and international collaboration. He was the lead author of the first national HIT strategic plan, which was released in June 2008, and he led the development of a memorandum of understanding on eHealth between the European Union and the United States.

Prior to his work in the government, Dr. Friedman was associate vice chancellor for biomedical informatics and the founding director of the Center for Biomedical Informatics at the University of Pittsburgh. Prior to his time in Pittsburgh, he served in a range of faculty and administrative roles in the School of Medicine at the University of North Carolina at Chapel Hill.

Dr. Friedman is an elected fellow and past president of the American College of Medical Informatics and an associate editor of the *Journal of the American Medical Informatics Association*. He was the 2011 recipient of the Donald Detmer award for policy innovation in biomedical informatics.

Geoffrey Ginsburg, M.D., Ph.D., is the founding director for the Center for Applied Genomics at the Duke University Medical Center and the founding executive director of the Center for Personalized and Precision Medicine in the Duke University Health System. He is a professor of medicine, pathology, and biomedical engineering at Duke University. He is an internationally recognized expert in genomics and personalized medicine with funding from the National Institutes of Health, the U.S. Department of Defense, the U.S. Air Force, the Defense Advanced Research Projects Agency, the Gates Foundation, and industry. Prior to working at Duke he was at Millennium Pharmaceuticals Inc., where he was vice president of molecular and personalized medicine and was responsible for developing pharmacogenomic and biomarker strategies for therapeutics.

He serves as an expert panel member for Genome Canada, as a member of the Board of External Experts for the National Heart, Lung, and Blood Institute, as co-chair of the Institute of Medicine's Roundtable on Genomic-Based Research for Health, as a member of the advisory council for the National Center for Accelerating Translational Science, as co-chair of the Cures Acceleration Network, as an advisor to the Pharmacogenetics Research Network, and as a member of the World Economics Forum's Global Agenda Council on the Future of the Health Sector.

Colin Hill, M.S., co-founded GNS Healthcare in 2000 and has served as its chief executive officer since then. He brings years of hands-on scientific experience to his role, with expertise in the areas of computational physics, systems biology, and personalized medicine. He also serves as chairman of Via Science, a leading big data analytics company focused on business intelligence, finance, and economic forecasting. In addition, Colin served on the board of directors of AesRx, a biopharmaceutical company dedicated to the development of new treatment for sickle cell disease. Mr. Hill was the founding chairman of O'Reilly Media's Strata Rx, one of the first health care big data conferences.

In 2004, Mr. Hill was named to Massachusetts Institute of Technology's TR100 *Technology Review* list of the top innovators in the world under the age of 35. He is a frequent speaker at national and international scientific and industry conferences and has been quoted in numerous publications and television programs, including the *Wall Street Journal*, CNBC Morning Call, *Nature*, *Forbes*, *Wired*, and *The Economist*. He graduated from Virginia Tech with a degree in physics and earned master's degrees in physics from both McGill University and Cornell University.

Gail Jarvik, M.D., Ph.D., received her Ph.D. at the University of Michigan and her M.D. at the University of Iowa in the medical scientist training program. She completed her residency in internal medicine at University of Pennsylvania and a fellowship in medical genetics at the University of Washington. Dr. Jarvik holds the Arno G. Motulsky Endowed Chair in Medicine and heads the Division of Medical Genetics and the Northwest Institute of Genetic Medicine. She cares for adult medical genetics patients. Her research focuses on the genetic basis of common diseases and the implementation of genomic medicine. Additionally, she has active research in biomedical ethics, including returning genomic research results to subjects. She is a principal investigator in the Electronic Medical Records and Genomics, or eMERGE, and the Clinical Sequencing Exploratory Research consortia.

Andrew Kasarskis, Ph.D., is the vice chair of the Department of Genetics and Genomic Sciences and the co-director of the Icahn Institute for Genomics and Multiscale Biology at Mount Sinai Hospital. His research is focused on developing and applying technology to biological problems including pathogen surveillance, pharmacogenomics, and the genetics of sleep. Prior to working at Mount Sinai he held positions at Pacific Biosciences, Sage Bionetworks, and Merck, and he has more than a decade of experience managing research and technology development projects in software engineering, drug development, human and mouse genetics, and other applications of biological research. Dr. Kasarskis holds a Ph.D. in molecular and cellular biology from the University of California, Berkeley, as well as a B.S. in biology and a B.A. in chemistry from the University of Kentucky.

Stephen Leffler, M.D., is the chief medical officer at The University of Vermont Medical Center and a professor of surgery at the University of Vermont College of Medicine. Appointed in 2011, he is responsible for medical staff affairs, the Jeffords Institute for Quality, technology management, and ethics, and he also participates in strategic planning and leads continued development of a regional integrated system of care, drawing on his experience working with other hospitals and physicians in the area. Dr. Leffler served as the medical director of the emergency department at The University of Vermont Medical Center until October 2011 and served as president of the medical staff at The University of Vermont Medical Center from 2010 to 2011. He has served on numerous clinical committees during his more than two decades as an emergency medicine physician, and he has been a key collaborator on significant organizational initiatives, including The University of Vermont Medical Center's regional ST elevation myocardial infarction project, an innovative program to ensure heart attack victims receive lifesaving care as rapidly as possible. A past president of the Vermont Chapter of the American College of Emergency Physicians, Dr. Leffler received his medical degree from the University of Vermont College of Medicine and completed residency training in emergency medicine at the University of New Mexico before joining the University of Vermont/Fletcher Allen faculty in 1993.

Debra G. B. Leonard, M.D., Ph.D., is a professor and the chair of the Department of Pathology and Laboratory Medicine at The University of Vermont Medical Center in Burlington, Vermont. She is an expert in the molecular pathology of genetic diseases, cancer and infectious diseases, and policy development for genomic medicine. Her M.D. and Ph.D. degrees were completed at the New York University School of Medicine, where she also did her postgraduate clinical training in anatomic pathology, including a surgical pathology fellowship. She is certified by the American Board of Pathology in anatomic pathology and by the American Boards of Pathology and Medical Genetics in molecular genetic pathology. Currently, Dr. Leonard is a member of the Institute of Medicine (IOM) Roundtable on Translating Genomic-Based Research for Health, and she previously served as a member of the IOM Committee on the Review of Genomics-Based Tests for Predicting Outcomes in Clinical Trials. She is a fellow of the College of American Pathologists (CAP) and the chair of CAP's Personalized Healthcare Committee. Dr. Leonard is a past member of the Secretary's Advisory Committee on Genetics Health and Society to Secretary Michael O. Leavitt and a past president and 2009 Leadership Award recipient from the Association for Molecular Pathology. She has spoken widely on various molecular pathology test services, the future of molecular pathology, the impact of gene patents on molecular pathology, and the practice of genomic medicine.

Scott Moss leads the research informatics research-and-development team at Epic. His team focuses on designing and developing research-enabling functionality within Epic's software, including the integration of genomic data into the electronic health record. In addition, Scott has been involved in several industry standards development initiatives, including participation in Integrating the Healthcare Enterprise and Health Level 7.

John David Larkin Nolen, M.D., Ph.D., M.S.P.H., is the managing director of laboratory medicine at Cerner Corporation. He earned his medical degree and his doctorate and his master's degree in public health and biomedical engineering and his undergraduate degrees in electrical and mechanical engineering from Tulane University in New Orleans, Louisiana. Before coming to Cerner, Dr. Nolen served as the assistant medical director at CSI Laboratories in Alpharetta, Georgia, where he provided medical oversight and diagnostic interpretation in cancer diagnostics. Dr. Nolen also served as chief medical officer for LifeSouth Community Blood Centers in Gainesville, Florida, where he oversaw a large tri-state community blood center. Dr. Nolen completed his clinical pathology residency at Emory University in Atlanta, Georgia. Dr. Nolen completed fellowships in transfusion medicine and hematopathology at the University of Iowa in Iowa City.

Alexander Ommaya, D.Sc., M.A., is the senior director of implementation research and policy at the Association of American Medical Colleges (AAMC). In this role he is responsible for enhancing AAMC member impact and capacity in effectiveness and implementation research. Previously as director of translational research at the Department of Veterans Affairs, he was responsible for managing the development of new research activities focusing on health systems improvements, genomic medicine, comparative effectiveness research, and traumatic brain injury. Previously he was director of the Institute of Medicine's Drug Forum and Clinical Research Roundtable. These multi-stakeholder groups addressed science policy issues concerning the research enterprise and established public–private collaborative research activities. At Blue Cross and Blue Shield of Florida, he directed business knowledge management where his department evaluated and developed improvements for pharmacy, disease, and utilization management programs. Dr. Ommaya has worked for the Agency for Healthcare Research and Quality as a senior advisor for the Office of the Administrator; for the Walter Reed Army Medical Center as a senior researcher for the Defense and Veterans' Brain Injury Program; for the U.S. Senate as a health policy fellow; and for the National Institute of Mental Health. His previous research focused on neuroplasticity and cortical reorganization; the treatment of malignant glioma; rehabilitation after traumatic brain injury; and health system structures that improve research translation and implementation. Dr. Ommaya received his doctoral degree in health policy and management from Johns Hopkins University, a master's degree in biopsychology from Mount Holyoke College, and his undergraduate degree in philosophy from Vassar College.

Ketan Paranjape, M.S., M.B.A., is the worldwide director of health and life sciences in the Health Strategy and Solutions Group at Intel Corporation. His role involves managing and partnering with a worldwide cross-Intel team from various business units, sales and marketing, global public policy, corporate affairs, and Intel Labs to develop health information technology strategies and solutions for health and life sciences customers, managing and driving results for multiple, simultaneous projects inside and outside of Intel that cover a wide range of topics from technology assessment to solution blueprints to public policy to innovation processes. With a background in data sciences and bioinformatics, he currently runs a corporate wide Big Data Analytics in Health and Life Sciences program where he is partnering with cross-Intel teams, academic medical centers, original equipment manufacturers, original design manufacturers, and independent software vendors to develop, market, and research specific solutions for payers, providers, governments, life sciences, and pharmaceuticals research and development with the goal of delivering "personalized medicine at the touch of a button . . . everywhere . . . everyday . . . and for everyone."

In his 17 years at Intel Corp, he has been part of products spanning the entire "compute continuum" from high-performance computing to embedded medical devices in roles encompassing architecture, research and development, strategy, planning, and sales and marketing. He has spent 3 years in Intel Research as the chief of staff and technical advisor to the Intel chief technology officer. He also had a stint at the International Telecommunications Union leading a high-level experts' group to create technical and procedural specifications for cybersecurity. He is currently a faculty member at the International Institute for Analytics and teaches regularly at the Harvard School of Public Health.

Prior to working at Intel, Ketan worked at Medtronics on implantable devices and at Fujitsu on hospital information systems, and he completed internships at Honeywell. He has received M.S. degrees in electrical engineering and computer sciences from the University of Wisconsin–Madison, and his M.B.A. from the University of Oregon. He recently completed a certificate program in leadership strategies for information technology in health care from the Harvard School of Public Health. Ketan has co-authored the book *Design of Pulse Oximeters*.

Josh Peterson, M.D., M.P.H., is an assistant professor of biomedical informatics and medicine within the Vanderbilt University School of Medicine. Dr. Peterson completed his M.D. degree at Vanderbilt University School of Medicine (1997), an internal medicine residency at Duke University Medical Center (2000), a fellowship in general internal medicine at the Brigham and Women's Hospital, and a master's of public health degree at the Harvard School of Public Health (2002). He practices internal medicine at the Vanderbilt Adult Primary Care Clinic.

As a dual appointee within biomedical informatics and internal medicine, Dr. Peterson has played a pivotal role in developing and evaluating clinical decision support to improve drug safety and efficacy and to translate genomic technologies to routine clinical care. He directs the development and evaluation of clinical decision support for patients tested within Vanderbilt University Medical Center's large-scale pharmacogenomics quality improvement initiative, PREDICT (Pharmacogenomic Resource for Enhanced Decisions In Care and Treatment). He currently leads research funded by the National Heart, Lung, and Blood Institute, the National Human Genome Research Institute, the Centers for Disease Control and Prevention, and the Pharmacogenomics Research Network to evaluate the implementation of PREDICT, including both clinical and cost effectiveness.

Neil Risch, Ph.D., is the Lamond Family Foundation Distinguished Professor in Human Genetics, the director of the Institute for Human Genetics, and a professor and former chair of the Department of Epidemiology and Biostatistics at the University of California, San Francisco (UCSF). He is also adjunct investigator at the Kaiser Permanente Northern California Division of Research. Dr. Risch received his undergraduate training at the California Institute of Technology in mathematics and received his Ph.D. from the University of California, Los Angeles, in biomathematics with a minor in genetics. Prior to coming to UCSF in 2005, Dr. Risch held professorships at Columbia, Yale, and Stanford universities. Dr. Risch's research interests are in the areas of human genetics, genetic epidemiology, and statistical genetics, where he has published extensively. He is recognized as a highly cited researcher by the Institute for Scientific Information and is the recipient of the Curt Stern Award from the American Society of Human Genetics for his contributions to human genetics. He is an elected fellow of the American Association for the Advancement of Science, the California Academy of Sciences, and the Institute of Medicine of the National Academies. Risch is responsible for the identification of a number of genes underlying important medical conditions such as hemochromatosis and torsion dystonia. He is recognized for his novel statistical approaches to the genetic study of common, complex diseases, in particular genome-wide association studies. Currently, he is co-director of the Research Program on Genes, Environment and Health at the Kaiser Division of Research, and he is the joint principal investigator of the largest genome-wide study to date—of a cohort of 110,000 Kaiser members that is focusing on genetic and environmental factors influencing age-related disease and healthy aging.

Fred Sanfilippo, M.D., Ph.D., is the director of the Emory–Georgia Tech Healthcare Innovation Program, which accelerates innovation in health care services research, education, and delivery, and he is a professor of pathology and laboratory medicine at the Emory School of Medicine and of health policy and management at the Rollins School of Public Health. He has been a physician–scientist leader at Duke University as chief of immunopathology; at Johns Hopkins University as chair of pathology and research director of the Comprehensive Transplant Center; at The Ohio State University (OSU) as dean of the College of Medicine, executive dean and senior vice president for health sciences, and the chief executive officer of OSU Medical Center; and at Emory University as the chief executive officer of the Woodruff Health Science Center, chair of Emory Healthcare, and executive vice president for health sciences. At each institution he led organizational and culture changes resulting in novel interdisciplinary programs and improved academic, clinical, and financial performance. He has been an advisor to numerous universities, companies, and government agencies; the board chair of five nonprofit organizations; and the president of seven professional societies. He has published 250 articles, received more than $30 million in research funding and 3 patents, served on 13 editorial boards, mentored 33 graduate students and fellows, and given more than 200 invited talks. He received a B.A. and an M.S. in physics from the University of Pennsylvania, an M.D. and a Ph.D. in immunology from Duke University, and boards in anatomic, clinical, and immuno-pathology.

Sam Shekar, M.D., M.P.H., is the chief medical officer within Northrop Grumman's information systems sector. In this capacity he provides strategic clinical direction for the health division and serves as an adviser to health care and public health organizations, government agencies, and customers and partners on technology and policy issues in the medical and public health fields. Dr. Shekar also serves as director of the division's life sciences program, providing guidance and leadership for Northrop Grumman's bioinformatics-related programs and initiatives.

Prior to joining Northrop Grumman, Dr. Shekar served as a physician consultant. Previously, he spent more than 21 years as an officer with the U.S. Public Health Service. A former assistant surgeon general and rear admiral, Dr. Shekar has held executive-level health policy and management positions at the Office of the Assistant Secretary for Health, the National Institutes of Health, and the Health Resources and Services Administration (serving as director of the bureaus of primary health care and health professions).

Dr. Shekar has also worked as a medical officer at the Centers for Medicare and Medicaid Services (providing advice and guidance on

Medicare coverage and coding policy) and as a medical epidemiologist at the Centers for Disease Control and Prevention. During the record-breaking 2005 hurricane season, he served in Louisiana and Florida, directing the federal health response in Florida to Hurricane Wilma.

Dr. Shekar is a board-certified fellow of the American College of Preventive Medicine and holds a medical license in the state of Maryland. In addition, he is a member of the Institute of Medicine's Roundtable on Translating Genomic-Based Research for Health and co-chair of its health care systems–focused committee. He earned bachelor's, master's of public health, and doctorate of medicine degrees from the University of Michigan in Ann Arbor.

Sharon Terry, M.A., is the president and chief executive officer of the Genetic Alliance, a network of more than 10,000 organizations, 1,200 of which are disease advocacy organizations. The Genetic Alliance improves health through the authentic engagement of communities and individuals. It develops innovative solutions through novel partnerships, connecting consumers to smart services.

She is the founding chief executive officer of PXE International, a research advocacy organization for the genetic condition pseudoxanthoma elasticum (PXE). As co-discoverer of the gene associated with PXE, she holds the patent for ABCC6 and has assigned her rights to the foundation. She developed a diagnostic test and is conducting clinical trials.

Ms. Terry is also a co-founder of the Genetic Alliance Registry and Biobank. She is the author of more than 90 peer-reviewed articles. In her focus at the forefront of consumer participation in genetics research, services, and policy, she serves in a leadership role on many of the major international and national organizations, including the Institute of Medicine (IOM) Board on Health Sciences Policy, the National Coalition for Health Professional Education in Genetics board, and the International Rare Disease Research Consortium Interim Executive Committee, and she is co-chair of the IOM Roundtable on Translating Genomic-Based Research for Health. She is on the editorial boards of several journals. She was instrumental in the passage of the Genetic Information Nondiscrimination Act. She received an honorary doctorate in 2005 from Iona College for her work in community engagement, the first Patient Service Award from the University of North Carolina Institute for Pharmacogenomics and Individualized Therapy in 2007, the Research!America Distinguished Organization Advocacy Award in 2009, and the Clinical

Research Forum and Foundation's Annual Award for Leadership in Public Advocacy in 2011. She is an Ashoka Fellow.

Jason Vassy, M.D., M.P.H., is a primary care physician and clinician–investigator at Harvard Medical School, the Veterans Affairs Boston Healthcare System, and Brigham and Women's Hospital. His research examines the clinical translation of genomic medicine to primary care settings, with a focus on how primary care physicians will use genomic technology in their medical decision making. He is also a co-investigator on the MedSeq Project, the first randomized trial of whole-genome sequencing in primary care. In this study, one of the National Institutes of Health Clinical Sequencing Exploratory Research (CSER) projects, Dr. Vassy is observing how primary care physicians receive and interpret the sequencing results of their generally healthy patients and how they discuss the results with their patients. This work will inform how health systems can support and optimize physician use of genomic medicine.

C

Statement of Task

An ad hoc planning committee will organize and conduct a public workshop to examine how various systems are capturing and making use of genomic data to advance patient care and research. The workshop goal will be to evaluate the challenges, opportunities, and best practices for translating genomic data into knowledge that can inform both basic research and clinical care. In this context, various sources, tools, and methodology for the assessment of data related to the health care system may be considered. A diverse stakeholder group which may be composed of electronic health record developers and health information technology professionals, clinical providers, academic researchers, patient groups, and government representatives will be invited to present their perspectives. The planning committee will develop the workshop agenda, select speakers and discussants, and moderate the discussions. An individually authored summary of the workshop will be prepared by a designated rapporteur in accordance with institutional policy and procedures.

D

Registered Attendees

Anjene Addington
National Institute of Mental Health
National Institutes of Health

Carol Alter
AstraZeneca

Hilary Andreff
Genetic Alliance

Naomi Aronson
Blue Cross Blue Shield Association

Samuel Aronson
Partners HealthCare Personalized Medicine

Mohammed Asiri
King Abudliziz City for Science and Technology

Adam Aten
Brookings Institution

Jason Aulds
Department of Defense

David Bachinsky
Molecular Creativity

Dixie Baker
Martin, Blanck & Associates

Suzanne Bakken
Columbia University

Paul Billings
Thermo Fisher Scientific

Katherine Blizinsky
National Human Genome Research Institute
National Institutes of Health

Meryl Bloomrosen
TJG

Bruce Blumberg
Kaiser Permanente Northern California

Richard Bookman
University of Miami

Angie Botto-van Bemden
Musculoskeletal Research
  International

Kimberly Boucher
Inova Health System

Khaled Bouri
U.S. Food and Drug
  Administration

Erica Breslau
National Cancer Institute
National Institutes of Health

Joel Brill
Predictive Health, LLC

Brenda Brooks
U.S. Food and Drug
  Administration

Philip J. Brooks
National Center for Advancing
  Translational Sciences
National Institutes of Health

Suanna Bruinooge
American Society of Clinical
  Oncology

Nessa Bryce
Quest University Canada

John Burch
JLB Associates

Tara Burke
Association of Molecular
  Pathology

Leah Burns
Bristol-Myers Squibb

Robert Campbell
Brown University

Alexis Carlson
Georgetown University
  Program for Regulatory
  Science and Medicine; Center
  of Excellance in Regulatory
  Science and Innovation
  (PRSM/CERSI)

Lucy Carruth
Johns Hopkins Applied Physics
  Laboratory

Sarah Carter
J. Craig Venter Institute

Ann Cashion
National Institute of Nursing
  Research
National Institutes of Health

Christopher Chute
Mayo Clinic

Michael Clare-Salzler
University of Florida

Gina Clemons
Social Security Administration

APPENDIX D

Elaine Collier
National Center for Advancing
 Translational Sciences
National Institutes of Health

Mick Correll
GenoSpace

Anurupa Dev
Association of American
 Medical Colleges

Safiyya Dharssi
Pfizer

Michael Dougherty
American Society of Human
 Genetics

Jennifer Dreyfus
Dreyfus Consulting, LLC

Kelly Dunham
Patient-Centered Outcomes
 Research Institute

Emily Edelman
The Jackson Laboratory

Lynn Etheredge
Rapid Learning Project

W. Gregory Feero
JAMA

J. Michael Fitzmaurice
JMF Associates

David Flannery
American College of Medical
 Genetics and Genomics

Scott Fogerty
Transomics Consulting

Matthew Foster
Georgetown University

Tom Fowler
Genomics England

Charles Friedman
University of Michigan Medical
 School

Liz Geltman
Hunter College, City University
 of New York School of Public
 Health

Geoffrey Ginsburg
Duke University

Michael Goldstein
Social Security Administration

Peter Goodhand
Global Alliance for Genomics
 and Health

Eric Green
National Human Genome
 Research Institute
National Institutes of Health

Jennifer Hall
Lillehei Heart Institute
University of Minnesota

Gregory Hess
University of Pennsylvania

Colin Hill
GNS Healthcare

John Greg Howe
Yale University School of
    Medicine

Pawan Jain
U.S. Food and Drug
    Administration

Gail Jarvik
University of Washington

Janet Jenkins-Showalter
Genentech

Brett Johnson
One Million Solutions in Health

Nicole Johnson
Invitae

AJ Jones II
The Podesta Group

Andrew Kasarskis
Icahn School of Medicine at
    Mount Sinai

Marcia Kean
Feinstein Kean Healthcare

Monisha Khan
Inova

Muin Khoury
Centers for Disease Control and
    Prevention

Jennifer King
American Society of Clinical
    Oncology

Lisa Klein
Inova Translational Medicine
    Institute

Raluca Kurz
University of California, Los
    Angeles
Fielding School of Public
    Health

Vincent Lau
American Association of
    Colleges of Pharmacy

Gabriela Lavezzari
PhRMA

Stephen Leffler
University of Vermont Medical
    Center

Debra Leonard
College of American
    Pathologists
University of Vermont Medical
    Center

Shira Levy
Inova Health System

Rongling Li
National Institutes of Health

Nita Limdi
University of Alabama at
  Birmingham

Mingkuan Lin
George Mason University

Robert Lipsky
Inova Health System

Suzanne Luther
Social Security Administration

Erika Lutins
Genetic Alliance

Julie Lynch
Veterans Health Administration

Joshua Mann
American Society of Clinical
  Oncology

Nicholas Marko
Geisinger Medical Center

Myles Maxfield
Mathematica

Robert McCormack
Johnson & Johnson

Kathleen McCormick
SciMind, LLC

Timothy Morck
Nestlé Corporate Affairs

Jennifer Moser
Veterans Health Administration

Scott Moss
Epic

Suman Mukherjee
Genetic Alliance

Sumitra Muralidhar
Department of Veterans Affairs

Nancy Myers
Catalyst Healthcare Consulting

John David Nolen
Cerner Corporation

James O'Leary
Genetic Alliance

Alexander Ommaya
Association of American
  Medical Colleges

Jennifer Pacheco
Northwestern University

Ketan Paranjape
Intel

Erin Payne
Northrop Grumman Health
  Division

Susan Pearson
Yakima Valley Farm Workers
  Clinic

Michelle Penny
Eli Lilly and Company

Josh Peterson
Vanderbilt University Medical
   Center

Robert M. Plenge
Merck Research Laboratories

Victoria Pratt
Association for Molecular
   Pathology

Daryl Pritchard
Personalized Medicine
   Coalition

Ronald Przygodzki
Department of Veterans Affairs

Ximena Qadir
National Cancer Institute
National Institutes of Health

Nalini Raghavachari
National Institute on Aging
National Institutes of Health

Sara Reardon
Nature

Kate Reed
The Jackson Laboratory

Neil Risch
University of California, San
   Francisco

Samantha Roberts
Friends of Cancer Research

Laura Rodriguez
National Human Genome
   Research Institute
National Institutes of Health

Helena Rubinstein
Freelance Journalist

Fred Sanfilippo
Emory University

Fatoumata Sangare
George Washington University
   Alumni

Derek Scholes
National Human Genome
   Research Institute
National Institutes of Health

Sheri Schully
National Cancer Institute
National Institutes of Health

Joan Scott
Health Resources and Services
   Administration

Regina Searcy
Searcy Venture Group

Shahid Shah
HITSphere.com

Lalitha Shankar
Cancer Imaging Program

Sam Shekar
Northrop Grumman

Jared Stevenson
Genetic Alliance

Brad Strock
Epic

Margaret Sutherland
National Institute of
  Neurological Disorders and
  Stroke
National Institutes of Health

Devin Sutton
McGirr CO KG

Katie Johansen Taber
American Medical Association

Sharon Terry
Genetic Alliance

Zivana Tezak
U.S. Food and Drug
  Administration

Quynh Tran
Cystic Fibrosis Foundation

Patty Vasalos
College of American
  Pathologists

Jason Vassy
Harvard Medical School
Veterans Administration
Boston Healthcare System

David Veenstra
University of Washington

Wes Walker
Cerner

Michael Watson
American College of Medical
  Genetics and Genomics

Meredith Weaver
American College of Medical
  Genetics and Genomics

Jennifer Weisman
Strategic Analysis, Inc.

Angele White
Fair Chance, Inc.

Catherine Wicklund
National Society of Genetic
  Counselors

David Wierz
OCI Group, LLC

Bob Wildin
National Human Genome
  Research Institute
National Institutes of Health

Janet Williams
University of Iowa and
  American Academy of
  Nursing

Catherine Witkop
U.S. Air Force

Andrew Womack
Genentech

Janet Woollen
Columbia University